Rocher

AF737610

Household Energy Use & Conservation

Household Energy Use & Conservation

How to Prepare an Energy Budget

John Luetzelschwab

Illustrations by C. Dewitt Long

Nelson Hall
nh
Chicago

Library of Congress Cataloging in Publication Data

Luetzelschwab, John.
Household energy use & conservation.

Bibliography: p.
Includes index.
1. Dwellings—Energy conservation. 2. Dwellings—Energy consumption. I. Title.
TJ163.5.D86L83 644 79-16895
ISBN 0-88229-476-8 (cloth)
ISBN 0-88229-773-3 (paper)

Copyright © 1980 by John Luetzelschwab

Reprinted 1980

All rights reserved. No part of this book may be reproduced in any form without permission in writing from the publisher, except by a reviewer who wishes to quote brief passages in connection with a review written for broadcast or for inclusion in a magazine or newspaper. For information address Nelson-Hall Inc., Publishers, 111 North Canal Street, Chicago, Illinois 60606.

Manufactured in the United States of America

10 9 8 7 6 5 4 3 2

To Marcia, Dana, and Mark

Contents

List of Abbreviations

Btu:	British thermal unit
Btuh:	Btu per hour
Ccf:	100 cubic feet
cf:	cubic feet
cfh:	cubic feet per hour
Ddh:	Degree-difference-hour
DEER:	Dehumidifier Energy Efficiency Ratio
EER:	Energy Efficiency Ratio
°F.:	Degree Fahrenheit
ft^2:	square feet
hr:	hour
kw:	kilowatt
kwh:	kilowatt hour
lb.:	pound

Preface

For many years, energy suppliers urged the public to use energy—and why not? It was cheap, plentiful, and the key to a rising standard of living. Being good Americans, we responded to the call in glorious fashion, developing a large appetite for electricity, natural gas, and petroleum. We used energy for any labor-saving device we could think of, from electric toothbrushes to big cars. We also used it wastefully; energy was cheap, so why expend money and effort to conserve it? If the cost of natural gas needed to heat a house was cheaper than the cost of extra insulation, it was economical for a house to have little insulation and use a lot of gas.

Then came the oil embargo of 1973. I, like many others, was faced with the possibility of having too little oil to heat our house. Those who use natural gas faced the same possibility during the winter of 1977. We heard cries of conserve, conserve, conserve. In 1973 I turned down our thermostat and we wore a lot of sweaters. The next year I added more insulation to my attic. But what prompted this action? Partially it was a response to the "energy crisis," but my efforts were mainly responsive to the increasing cost of heating oil. The rising cost of energy has created a powerful incentive to conserve.

Newspapers, magazines, and the radio have presented numerous energy conservation tips. But which conservation measures would really save enough energy to be worth the investment of time and money? While some conservation measures can save one family much energy and money, they may save another family only a little.

The problem of where to conserve and how much to spend on conservation measures is similar to that of balancing the family budget. If you have trouble balancing your budget, you first determine what you spend for various items: food, clothing, car, and so on; then you decide where to reduce that spending. You must follow the same process where energy is concerned. First find where you use energy and how much you use for your different appliances, then decide where to to reduce that energy use.

A simple accounting of your energy use can have beneficial results. When the management of a major corporation asked the department supervisors to account for their energy uses (even though not asking for any conservation measures), the energy consumption of most departments decreased by about 20 percent. Just the awareness of where the energy goes helps one to see where waste exists and where consumption can be reduced.

But energy conservation cannot be a momentary fling, a fad, or something we do only for a while. It must become a way of life in order to be effective. The supplies are decreasing and the cost is increasing. Our children must learn proper conservation habits if they are to maintain the same standard of living throughout their lifetimes. The greatest legacy we can leave them is an understanding of how to reduce the energy waste in their lives and of how to live without so many energy-eating appliances. I am now thankful to my own parents for their constant reminders to turn off the lights, close the door, turn off the hot water, and so on. And I now have a greater incentive to correct my own children's bad energy habits; they need to be prepared for a world that may lack abundant energy.

Although this book is designed primarily for use by the homeowner, it can also serve as a text for high school, college, or adult night school courses on energy conservation. Even a home economics course might find it useful.

I would like to extend my sincere gratitude to several people who have generously given their time for this book. First I wish to thank Dr. David Telfair of Earlham College, my father Edgar J. Luetzelschwab, and Robert Romancheck and Harold Nelson of Pennsylvania Power and Light Company for thorough reading of the manuscript and helpful suggestions. Thanks also go to my colleagues Ken and Priscilla Laws, Bill Deutschman, and Neil Wolf for their comments relating to some of the physics in the book. I also thank Drue Sherman for providing a layman's point of view of the text. But my biggest thanks go to my typist Martha Wilson; without her, this book would still be only a pile of paper filled with my scribbling.

How to Use This Book

Most books you just pick up and read, then put down when you're finished. This book is a little different. If you want to gain fullest use of this book, you'll have to do some work yourself.

This book has two goals: to allow you to calculate the amounts of energy used in your home; and to help you understand the concept of energy and how your home appliances use that energy. If you only follow the energy conservation measures listed in this book or other sources, you may remain unsure of just how those actions help save energy. But if you understand the concepts of energy and know how much you pay for it, you will gain a better idea of how much each conservation measure can save you.

Because the calculations in a few sections are long and involve some effort on your part (measuring temperatures, areas, and so on), you should first read through the book to gain an understanding of what you can do with the calculations. A first reading will also increase your understanding of the energy concepts involved. Then you can decide which calculations you want to do. Although you might best start at the beginning and proceed through the calculations in order, this approach is not necessary. You can start with a later section, such as kitchen appliances—

but in order to calculate the cost of the energy used and the amount of energy saved, you'll need to complete the initial calculations for the cost of your energy (chapter 2; tables C-1 to C-6, Appendix C). You can work out the calculations at your leisure; do the ones that interest you when you feel like doing them. All of the calculation tables are found in Appendix C. The tables in the book are designed to make the calculations as simple and straightforward as possible.

Finding out where your energy goes and what it costs will require you to take a few measurements and gather some firsthand information of your own. In order to gather these few facts about your home heating system, your hot-water system, and a few of your electrical appliances, you'll need the following highly scientific pieces of equipment: a ruler or tape measure, a quart or gallon container, a watch or clock with a second hand, and a thermometer. The thermometer should read from 40° to about 150° F. If you can't obtain a thermometer that reads up to 150° F, use a cooking thermometer to measure the higher temperatures. Then, of course, you'll need some paper and a pencil or pen to complete the suggested calculations.

Now what about these calculations? If you're good at mathematics and enjoy adding, subtracting, multiplying, and dividing numbers, your pencil and paper should be enough. But if handling numbers frustrates you, maybe investment in a small calculator may make life a little easier. The amount of energy used by a rechargeable calculator is small—and if use of a calculator means that you'll make a more dedicated effort to save energy, the amount of energy saved will not only be considerably more than that used by the calculator—the money saved should be enough to pay for many calculators. If you use a calculator, you may want to refer to Appendix B to understand significant figures.

Each appliance section of the text contains a short description of that appliance, how it operates, one or more methods you can use to determine the amount of energy used by that appliance, and several actions you can take to save energy. Sometimes, depending on the information available to you, you can use more than one method to calculate the amount of energy used by an appliance. In these cases, the easiest method appears first and the more difficult methods follow. Often the more difficult methods are more accurate than the easier methods, so you may have to decide how accurately you want to measure your energy use or how much effort you want to put into the measurement. Some sections also contain different methods for different types of appliances: electric, oil, or gas. In these cases, read only the section that pertains to your type of appliance.

The best way to proceed through the calculations is to treat your energy consumption as you would treat your family budget. First determine how much energy you use for each appliance. As you proceed through the book, you can keep a tally of the energy use in table C-117 (Appendix C). After you have clearly defined your present uses of energy, you can see where most of it is going; then you can institute as many energy-saving measures as you want and determine how much you should save.

Sections describing the various conservation measures will present methods for calculating the amounts of energy saved by certain actions. You can keep a tally of these expected savings in table C-118 (Appendix C). For other actions, however, the data needed to calculate the energy savings are difficult or impossible to acquire, and therefore no calculation of energy savings is possible.

When keeping a tally of your expected energy savings, you should be aware of the relations between appliance use and heat in the house. Some energy actions actually may not save any energy during the heating season. You may want to refer to chapter 9, the section "When Is a Kilowatt Hour Saved Not a Kilowatt Hour Saved?" to see under what conditions some energy savings actions may be a futile effort.

1.

Introduction

Because of decreasing supplies and increasing costs of energy, people are making efforts to conserve it. One group of experts, the Henry Elementary School students in St. Louis, Missouri, offered the following energy saving measures:[1]

In Winter:

Lower people's body temperature to 68° F.
Make it a rule that there has to be at least two people in every bed that uses an electric blanket.
When beds aren't in use take the blankets and put them around the room to hold the heat.
Put more hot sauce in the food.
Don't have so many days of school.

In Summer:

Don't drink so much water; then you won't have so much humidity in you, and you won't feel the heat so bad.
Let birds fly around the house to keep the air circulating.

Anytime:

Eat more carrots so we can see just as good with less light.
Dip everything that's made in stuff that glows in the dark.
Don't stay in more than one room at a time.

Unfortunately, saving energy isn't quite that simple.

Household appliances are so much a part of our everyday life that the only time we really become conscious of them is when the electricity goes off. We have become very complacent about buying any appliance to make life easier and expecting to have enough energy to operate it. With our decreasing energy supplies and the rising cost of energy, however, we need to examine our uses of appliances to determine which ones we can do without, and which ones we can use more efficiently.

Appliances, Energy Savings, and Inconvenience

What is an appliance? We know that it is something that always seems to cost more than we can afford; it seems to break down just after the guarantee runs out or just after it is paid for; and, as we are all learning these days, it requires expensive energy to operate. Basically a household appliance is something that performs work for us, gives us enjoyment, or makes us more comfortable. Under this definition, home heating and hot water systems are appliances as well as dishwashers, television sets, and vacuum cleaners.

With very few exceptions, an appliance costs as much or more to operate than its purchase price. A $400 refrigerator can consume $800 worth of electricity over its expected fifteen-year lifetime, and a $30 hair dryer can consume 30 to $60 worth of electricity over a fifteen-year period. Therefore the purchase price is only half or less of the appliance's total cost. When purchasing an appliance, one needs to consider the operating cost just as much as the purchase price. Only two major appliances cost more to purchase than to operate—color television sets and microwave ovens.

Though people are mystified by the operation of most appliances, the basic operation of all appliances is not as difficult to understand as one might expect. A clothes washer, for example, does no more to the clothes than the pioneers did when they washed their clothes in a kettle of hot, soapy water, rinsed them in clear water, and wrung them out by hand. A modern washing machine agitates the clothes in soapy water (just like stirring them in a kettle), then rinses and spins them dry; all you do is put in the clothes and soap, push a button or turn a dial, and return when the clothes are finished—no building a fire, no carrying the kettle and water, no stirring, rinsing, and wringing the clothes. But you pay a price for this convenience: the possibility that a complicated appliance may malfunction as well as the actual cost of the energy you pay to operate it.

So how can you coax this complicated assortment of nuts and bolts to use less energy? Talking to appliances does no good because they are poor listeners. Energy-saving tips appear frequently in newspapers, magazines, on radio and television—"turn off the dishwasher before the dry cycle," "take a shower rather than a bath," "repair leaky faucets," "increase ceiling insulation to six inches," and so on. But how much energy and money can you actually save by following these tips? Some of them can save a lot, some only a little, and a few actually result in more energy use. Also, one tip may save one family a lot of energy but save another very little, depending on the lifestyles of the two families. And what about the inconveniences these actions may cause? Are they worth the energy and money saved?

Some energy-saving tips list certain amounts of money that can be saved, but these amounts are based on average families owning average appliances that use average amounts of energy at some average cost. On the average, however, nobody is average. To determine how much energy and money *you* can save, *you* must calculate the energy used by *your* appliances and calculate what *you* pay for that energy.

This book has two purposes: to explain the basic operation of each appliance so that you can understand how it uses energy; and to show you how to measure the actual energy used by each of your appliances so you can calculate the operating cost of each appliance. Armed with this knowledge, you can then intelligently adjust your energy consumption in order to save the maximum amount of energy with minimum inconvenience and discomfort. Sometimes, however, the only way to save energy is to change your way of doing something. In these cases, you can decide if the money saved is worth the change and possible inconvenience. If you simply replace high-wattage light bulbs with low-wattage bulbs where a bright light is unnecessary, for example, energy will be saved with little or no inconvenience to yourself. On the other hand, if you decide to stop using your clothes dryer and hang your clothes outside to dry, the energy you save will be at the expense of your time and convenience.

But how much savings of energy and money are worth how much inconvenience? A little energy saved each day adds up to a surprising amount of energy and money over several years. Advertisements may state that a certain appliance has some convenience feature that costs only "pennies per day" to operate. But remember that appliances will operate for many days, so a few pennies per day becomes many pennies over the lifetime of the appliance. A convenience feature that costs only 1 penny per day will cost $3.65 per year or nearly $55 over the normal fifteen-year life of most appliances. When considering that a penny's worth of energy today could cost two or three pennies (or more) in ten or fifteen years, the cost could be considerably more.

We can save energy in two ways: either reduce waste or reduce the use of our appliances. The inconvenience resulting from an energy saving action obviously depends on whether we reduce waste or reduce use. Some waste reduction actions, such as remembering to turn off lights when leaving a room, require thought and action every time someone walks out the door, and a person who forgets to turn out a light is inconvenienced by the return trip to the room. But most waste reducing actions, such as replacing high-wattage light bulbs with low-wattage bulbs, requires thought and action only once (when the bulbs are changed); yet energy is saved, with no inconvenience to the user, every time the light is needed. These two types of actions could be labeled "conscious" and "nonconscious" actions. A conscious action requires frequent thought and action and therefore is an inconvenience. A nonconscious action requires no thought after the initial action and therefore is less of an inconvenience.

In many cases, one can reduce waste energy by intelligent buying. One model of an appliance may consume much more energy than another, even when both appliances do the same job. The cheaper model often uses more energy, although not always. Again, knowing how to calculate the energy used by an appliance is necessary for determining which model is more efficient and therefore cheaper to operate. Air conditioner efficiencies vary greatly—two models may give the same amount of cooling, but one may use twice as much energy as the other. And since air conditioners are large energy users, the money *saved* over the lifetime of an air conditioner by the more efficient unit can actually be greater than the purchase price of the appliance itself.

One can generally find ways to reduce waste and not change one's way of living. If you reduce your use of an appliance, however, you will need to change your lifestyle. If you quit using your clothes dryer, you must expend time and effort to hang up the laundry. Or if you turn down the thermostat during the winter, you must change your way of dress. But remember that any resulting inconvenience must be balanced against the benefits gained by the money saved and subsequently spent for other things. Will your inconvenience be worth the extra clothes you can buy, the extra food for your family, or the fewer headaches from trying to balance your budget?

Viewing the problem on a national scale, we might look into our crystal ball to see what the future may bring. Unfortunately our crystal ball doesn't give a clear picture. Through the mist we see the strong possibility of an economy no longer constantly expanding, rather it has reached a "steady state" because of limited resources of energy and materials. An important part of this future economy is conservation. In the "steady state" economy, people who have acquired energy conservation habits will enjoy a considerable advantage over those who have not.

2.

Energy Units: Costs and Uses

What is energy? Most physics books define energy as "the ability to do work." All of us like to have our work done for us—energy will wash our clothes and dishes, lift heavy loads, keep us cool in summer, cut our wood, and generally make life easier for us. Energy also comes in the form of heat, which may not seem like work but is actually a form of it. And pushing a button or turning a dial and letting the electricity or gas heat the oven is certainly a lot less work than chopping and burning wood for the same purpose.

How do you buy energy? When you purchase food at the grocery store, you pay for it by the pound, quart, piece, or some other convenient measure. Energy also comes in different measures. Electricity is sold in units of kilowatt hours (kwh), furnace oil in units of gallons, and natural gas in units of cubic feet—although what you may pay for is something called "therms," a measure of the actual energy in the gas.

A kilowatt hour is 1,000 watt hours (wh), whereas the watt hours used by an appliance is the wattage of the appliance multiplied by the time in hours that the appliance is using the electricity. For example, a 100-watt bulb that is lighted for 1 hour uses 100 watts × 1 hour = 100 wh, or 0.1 khr (100 wh ÷ 1,000 = 0.1 kwh); if lighted for 5 hours, the bulb would use 500 wh (100 watts × 5), or 0.5 kwh; and if on for 10 hours, it would use 1,000 wh, or 1 kwh. Note all times must be expressed in hours. For time less than an hour or times of hours and minutes, the minutes must be converted to a fraction of an hour. A 1,600-watt appliance used for 15 minutes (1/4 hour) consumes 1,600 watts × 1/4 hour = 400 wh = 0.4 kwh. (See Appendix A for further details on the use of units.)

The kilowatt hour is a measure of energy, but the gallon of oil and cubic foot of gas are measures of volume, not energy. Oil and gas are used entirely to produce heat, and therefore we measure the heat energy from them in the standard English unit of heat—the British thermal unit, or Btu. The Btu is defined as the energy needed to heat one pound of water by one degree Fahrenheit (1° F.) which is about the amount of energy given off by a single wooden match when burned. Now, who measures water by the pound? To gain some idea of a pound of water, remember the adage "a pint's a pound the world around"—a pint of water weighs approximately 1 lb. A gallon of water weighs about 8 lb. (8.34 to be exact), so to heat that gallon by 1° F., you need 8.34 Btu; to heat a gallon of water by 10° F., you need 83.4 Btu; or to heat 10 gallons of water by 1° F., you need 83.4 Btu. The number of Btu needed to heat water equals the weight of the water (in pounds) multiplied by the temperature rise of the water (in Fahrenheit degrees).

When burned, a gallon of furnace oil (#2 fuel oil) can produce about 140,000 Btu of heat, and a cubic foot of gas can produce about 1,000 Btu. Fuel oil generally gives about the same amount of heat per gallon regardless of its particular refinery source, but the energy per cubic foot of gas can vary somewhat depending on the well from which it originates. Therefore many local gas companies measure the actual heating value of the gas sent to their customers, then bill them for the actual energy delivered rather than for cubic feet of gas used. The unit of measure for energy from gas is the therm; a therm is defined as 100,000 Btu of heat energy. One cubic foot of gas can produce about 1,000 Btu, so about 100 cubic feet of gas are needed to provide 1 therm of heat. Because the heating value of gas doesn't vary by more than 5 percent from 1,000 Btu per cubic foot, we will consider 100 cubic feet and 1 therm of gas to be equivalent. This assumption avoids a lot of messy mathematics later. The standard abbreviation for 100 cubic feet of gas is Ccf, where the first C represents the Roman symbol for 100, and the second c and the f stand for cubic feet. Therefore, 100 cubic feet of gas = 1 Ccf = 1 therm = 100,000 Btu.

Although we use electricity to run motors for fans, compressors, and pumps, we also use it as a source of heat for toasters, ovens, clothes dryers, hot-water heaters, and the home. The electricity passes through a resistance heating element that becomes hot and then heats the air or water around it. In all cases where electricity is converted to heat, each kilowatt hour produces 3,413 Btu of heat.

Reading Your Gas and Electric Meters

Your electric and gas meters can play an important role in determining the energy used by the various appliances in your home. Since the utility companies use them to determine your energy use, you may as well use them too. First, of course, you must find your meter or meters. Usually the electric meter is on the outside of the house where the electrical lines enter; and the gas meter is located in the basement or outside where the gas line enters. If you have no basement, just go searching—the meter has to be somewhere accessible to the meter reader. If you live in an apartment, your electric and gas meters may be in the basement or service area.

The electric meter and the dials on the front should look like those in figure 1. To read the meter, start with the dial on the right. Each number represents 1 kwh, so a full turn measures 10 kwh. In figure 1, the pointer is close to the 7, so read it as 7. Next, look at the second dial from the right—each number represents 10 kwh. Find the *number it has passed;* that number is the tens of kilowatt hours. Be careful, this dial moves counterclockwise—you can tell by the counterclockwise arrangement of the numbers. This dial shows the pointer past the 9 but not yet to 0, so the reading for the two dials on the right is 97 kwh. The third dial from the right reads the hundreds of kilowatt hours; the pointer is almost but not quite on 5 because the second dial hasn't passed 0. Therefore the third dial reads 4. The pointer on the last dial (which, like the second dial, rotates counterclockwise) is past 6, so the meter reading is 6,497. Some meters have a fifth dial that reads up to 100,000 kwh, and some meters, especially in homes with central air conditioning or electric heat, have "X10"

Fig. 1. Electric meter

stamped on the right of the dials. For these meters, the entire reading must be multiplied by 10. If an X10 were present in figure 1, the reading would be 64,970 kwh.

To check yourself, write what you think the following meters read. Then check your answers with the correct readings given below figure 2.

Dials on a gas meter are read the same way as those on the electric meter and should look something like those shown in figure 3. The top two dials are test dials and can be used to measure small volumes of gas.

The three dials on the bottom indicate the cubic feet of gas used. Each number on the first dial on the right represents 100 cubic feet (1 Ccf); each number on the center dial represents 1,000 cubic feet; and each number on the far left dial represents 10,000 cubic feet. The meter in this example reads 42,800 cubic feet (428 on the dials × 100), or 428 Ccf.

Rates of Energy Use

To find the total amount of energy use by an appliance, you must know the rate of energy use (how much energy is used per time) and the amount of time the appliance is using energy. Rate of electrical energy consumption is measured in watts, and the rate of gas consumption is measured in cubic feet per hour, cubic feet per minute, or Btu per hour. Usually electrical appliances have the wattage listed on them, but gas appliances generally do not indicate their gas consumption rates. For your appliances that have no wattage or gas consumption listed, you can use your electric and gas meters to determine these values by the following methods.

Wattage Determination: Rotating Disc Method

Below the dials on your electric meter is a disc that rotates whenever electricity is flowing through the meter; this disc turns the pointers on the dials. Each turn of the disc measures a certain amount of energy (kilowatt hours) consumed, and the rate at which the disc turns is a measure of the total wattage of the appliances being used in your house. The faster the disc rotates, the higher the wattage of your appliances.

The energy measured by each turn of the disc is listed on the front of the meter as the Kh value (Kh is just a symbol—it has no

Fig. 2. Examples of electric meter dials. A: 7852 B: 5938 C: 4282

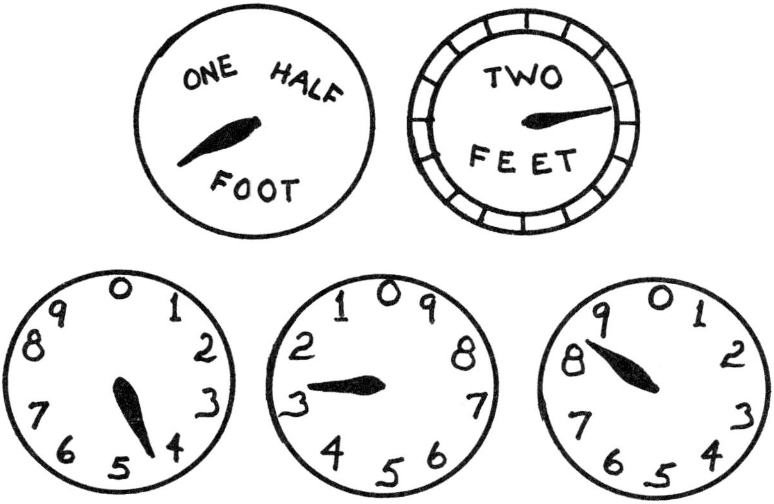

Fig. 3. Examples of gas meter dials

standard meaning). Most meters have Kh values of 1.8, 3.6, or 7.2 wh per revolution; that is, one revolution of the disc represents 1.8, 3.6, or 7.2 wh used. For future reference, look at the face of your meter, find the Kh value, and record that value on table C-1 (Appendix C).

To use the Kh value to measure the wattage of an appliance, we must consider the definition of watts and watt hours. As we have seen, the number of watt hours used by an appliance is the figure resulting from wattage multiplied by the time (in hours) the appliance is operating; so if your meter has a Kh value of 7.2 wh per revolution, a 7.2-watt appliance would cause the disc to make one revolution in 1 hour (7.2 watts × 1 hour = 7.2 wh). A 14.4-watt appliance would cause the disc to make one revolution in 1/2 hour (14.4 watts × 1/2 hour = 7.2 wh). Or considering the reverse situation, the wattage of an appliance is the Kh value divided by the time required for one revolution of the disc. For the 14.4-watt appliance, the wattage would be: 7.2 wh per revolution (Kh value) ÷ 1/2 hour = 14.4 watts.

Since most appliances have higher wattages than this, however, the meter disc usually makes one revolution in seconds rather than hours. Therefore to find the wattage of an appliance in your house, you must multiply the Kh value by 3,600 (number of seconds in 1 hour), then divide the resulting product by the number of seconds for one revolution.

To find the wattage of an appliance, use the following procedure:

1. Find a time when no appliances are in use. Don't bother to unplug the clocks—they don't use enough electricity to affect this experiment. You may want to turn off the refrigerator and freezer for the few minutes required to take these measurements, but don't forget to turn them on again when you're finished. If you have an instant-on TV, you should unplug that also.
2. After everything is turned off, check the disc on the meter to be sure it isn't turning, or is turning very slowly, then turn on the selected appliance. Then time one revolution of the disc. If the disc revolves once in about 15 seconds, measure the time it takes for two or three revolutions, and divide the total time by the number of revolutions to find the time for one revolution. The disc has a black indicator mark on the edge which you can use for timing. Start your timing when the mark appears at the front of the meter, and stop when it appears in the same position.
3. After measuring the time, turn off the appliance and check to see that the disc has stopped or is turning very slowly again. If the disc is still rotating, then another appliance is operating. In this case, disregard your previous numbers, check that all appliances are off again, and repeat the time measurement.
4. To find the wattage of the selected appliance, multiply the Kh value by 3,600 and divide that product by the number of seconds required for one revolution of the disc. For example, if the meter Kh value is 7.2 wh per revolution and the disc makes one revolution in 15 seconds, then the wattage of the appliance is 3,600 × 7.2 wh ÷ 15 seconds = 1,728 watts.

*Determination of Cubic Feet of Gas per Minute: Test Dial Method**

Test dials on a gas meter rotate as the gas flows through the meter. To measure the gas flow rate, all you need to know is the time required for the half-cubic-foot or the two-cubic-foot dial to make one revolution. The flow rate is the volume of gas (half-cubic-foot or two cubic feet) divided by the revolution time. For most gas consumptions to be calculated, you will use the half-cubic-foot dial.

To measure the gas consumption of an appliance, use the following procedure:

1. Select a time when no gas appliance is in use; check the half-cubic-foot test dial to see that the pointer is not rotating or rotating only slowly from gas for

*Although sections and tables are provided throughout this book for various fuels or different types of appliances, use only those that apply to your situation. If you don't use gas in your home, then ignore this section.

the pilot lights. Turn on the selected appliance and find the time required for the half-cubic-foot dial to make one revolution. Because the dial has no markings, you may want to mark a starting point on the dial with a pen or piece of tape. Divide 30 by the number of minutes required for one revolution of the pointer to find the cubic feet per hour (cfh) of gas flow for that appliance. Space is provided in Appendix C for the calculations. For example, if the gas flowing to a certain appliance causes the half-cubic-foot dial to rotate once in 4 minutes, then that gas appliance is using 30 ÷ 4 minutes = 7.5 cubic feet per hour (cfh) of gas flow.

2. In your later calculations for gas consumption by a gas appliance, you will need to make a correction for the gas flowing to pilot lights of all gas appliances in the house. To find the amount of gas consumed by these pilot lights, use the test dial method just outlined. If you have more than one gas appliance with a pilot light, you cannot determine the gas used by each pilot light without shutting off all but one at a time—but this practice is difficult and possibly dangerous. But you can easily determine the gas used by all the pilot lights and divide that total consumption by the number of pilot lights to find the average consumption of each.

Find a time when no gas appliance will be used so that the only gas being consumed is for the pilot lights. Preferably begin just after the hot water heater has been operating so that it won't come on for a while, and then be sure that nobody uses any hot water. With a pen or piece of tape, mark the position of the pointer on the half-cubic-foot dial on the gas meter, and find the time required for the pointer to make one complete revolution. If a gas appliance turns on during this time, begin again after everything is off. If you have only one appliance with a pilot light, the pointer may move very slowly, so time only a half or quarter revolution and multiply that time by 2 or 4 to find the time required for one revolution.

3. Record the time in table C-2 (Appendix C) and calculate the cubic feet used per hour. The amount of gas consumed by all pilot lights per month is the hourly consumption multiplied by 732 hours per average month (30.5 days), and the amount of gas consumed by each pilot light is the total consumption divided by the number of pilot lights.

Energy Costs

The cost of operating a given appliance not only depends on the amount of energy it uses but also on the cost of that energy. Tables and charts that show how much money an average family spends for the use of various appliances often appear in newspapers, magazines, and books. As previously mentioned, these figures are based on an average amount of energy at some average cost. But *your* energy costs are probably not the same as these average energy costs, so you must determine how much you actually pay for the energy *you* use. Although we will see some examples of costs and dollar savings later in this book, such examples are meant only as guides to show how to calculate your energy use and costs, or to indicate approximate costs; they should not be used as the actual cost of the energy you use.

Oil

Oil is sold by the gallon; although the price per gallon may vary with time and locality, the price is not dependent on the amount you use—you pay the same price per gallon for one hundred gallons as you do for two hundred gallons. List the price you pay for a gallon of oil in table C-3 (Appendix C) for future reference. If your cost has been rising, you may want to consult your heating oil dealer to get an approximate projected cost for oil during the next year or two. Record that value instead of your present cost in table C-3.

Electricity

Calculating the cost of electricity is not as simple as for oil. Electricity is sold by the kilowatt hour, which is simple enough—but the price per kilowatt hour decreases as you use more electricity each month and that makes things not so simple. An example of a residential rate schedule for one month's use is shown in table 1.

TABLE 1
Example of an Electrical Rate Schedule

Range	*Cost*
0–30 kwh	$2.75
Next 70 kwh	0.0452 (4.52 cents) per kwh
Next 400 kwh	0.0325 (3.25 cents) per kwh
Next 500 kwh	0.0275 (2.75 cents) per kwh
Over 1,000 kwh	0.0215 (2.15 cents) per kwh

If you are billed every two months rather than monthly, the above ranges are doubled.

Range	*Cost*
0–60 kwh	$5.50
Next 140 kwh	0.0452 (4.52 cents) per kwh
Next 800 kwh	0.0325 (3.25 cents) per kwh
Next 1,000 kwh	0.0275 (2.75 cents) per kwh
Over 2,000 kwh	0.0215 (2.15 cents) per kwh

TABLE 2
Example of an Electrical Rate Schedule with Fuel Adjustment Charges Added

Range	*Cost*
0–30 kwh	$3.04 ($2.75 + 30 kwh × $0.0095 per kwh)
Next 70 kwh	0.0547 per kwh ($0.0452 + $0.0095)
Next 400 kwh	0.0420 per kwh ($0.0325 + $0.0095)
Next 500 kwh	0.0370 per kwh ($0.0275 + $0.0095)
Over 1,000 kwh	0.0310 per kwh ($0.0215 + $0.0095)

If you have an electric hot-water heater or electrical heating in your home, your schedule may be different than the normal rate schedule. Your rates are probably a little lower than the normal residential rates, although some utilities have discontinued these lower rates or are discontinuing them for new customers.

In addition to these charges, your electric utility may add a fuel adjustment charge (FAC) that can range from almost nothing to a cent or two per kwh. This charge is a fixed cost per kwh and does not depend on your monthly consumption. The fuel adjustment charge can vary from month to month depending on the cost of the fuel that the electric utility buys. For example, with a fuel adjustment charge of $0.0095 per kwh, the total cost per kwh would be as shown in table 2 for the example rate schedule in table 1.

To show how the monthly costs for your electric bill are calculated, table 3 gives the calculation for a total of 600 kwh used per month. Your electric bill probably has the fuel adjustment charge listed as a separate charge, so we will use the rates listed in table 1 and add the fuel adjustment charge at the end.

Some states now have, and others will have, "lifeline rates."

TABLE 3
Sample Calculation of a Monthly Electrical Bill for 600 kwh

Range	*Cost*	
0–30 kwh	$ 2.75	
Next 70 kwh (30–100)	3.16	(70 kwh × $0.0452 per kwh)
Next 400 kwh (100–500)	13.00	(400 kwh × $0.0325) per kwh)
Next 100 kwh (500–600)	2.75	(100 kwh × $0.0275 per kwh)
	$21.66	
+ fuel adjustment charge	5.70	(600 kwh × $0.0095 per kwh)
	$27.36	

The philosophy of a lifeline rate is that a family should be able to purchase enough electricity for basic necessities of life at a reasonable cost but should pay a higher rate for any electricity in excess of this minimum amount. The rate for a lifeline quantity of electricity is less than the rate for any additional amount used over the lifeline quantity. The basic lifeline quantity is determined by the family's uses of electricity: basic use only (no electric hot-water heater or electric heat), basic use plus hot water, or basic use plus hot water and heat. An example of a lifeline rate schedule is shown in table 4.

Now, let's see what the monthly costs for a lifeline rate schedule would be for a family that uses 600 kwh per month with no electric heat or hot-water heater. The first 250 kwh costs 2.8 cents per kwh (lifeline rate, basic use quantity); the next 50 kwh left in the "first 300 kwh" category and over the lifeline quantity costs 3.0 cents per kwh; and the last 300 kwh costs 2.7 cents per kwh (nonlifeline rate over 300 kwh). If the family had an electric hot-water heater, the first 300 kwh would cost 2.8 cents per kwh; the next 200 kwh (the remaining amount in the 500 kwh lifeline quantity for basic use plus hot water) would cost 2.5 cents per kwh; and the last 100 kwh would cost 2.7 cents per kwh. (And you think your taxes are complicated!) The total monthly costs for these two situations are shown in table 5.

As if these cost calculations aren't complicated enough, many utilities have off-peak rates, charging less for electricity consumed from 8 A.M. to 8 P.M. This rate schedule requires a special electric meter. If you're curious, check with your electric utility.

TABLE 4
Example of a Lifeline Electrical Rate Schedule

Rates:	*Lifeline*	*Nonlifeline*
Customer charge	$1.75	$1.75
First 300 kwh	0.028 (2.8 cents)	0.030 (3.0 cents) per kwh
Over 300 kwh	0.025 (2.5 cents)	0.027 (2.7 cents) per kwh
Fuel adjustment charge (all kwh)	0.008 (0.8 cents)	0.008 (0.8 cents) per kwh

Lifeline Quantities per Month:	
Basic use	250 kwh
Basic use plus hot water	500 kwh
Basic use plus heat	800 kwh (winter), 250 kwh (summer)
Basic use plus heat and hot water	1050 kwh (winter), 500 kwh (summer)

TABLE 5
Sample Calculations for a Monthly Electric Bill for 600 kwh—Lifeline Rates

Basic use only:		
Customer charge	$ 1.75	
0–250 kwh	7.00	(250 kwh × $0.028 per kwh)
250–300 kwh	1.50	(50 kwh × $0.030 per kwh)
300–600 kwh	8.10	(300 kwh × $0.027 per kwh)
	$18.35	
+ fuel adjustment charge	4.80	(600 kwh × $0.008 per kwh)
	$23.15	
Basic use plus hot water:		
Customer charge	$ 1.75	
0–300 kwh	8.40	(300 kwh × $0.028 per kwh)
300–500 kwh	5.00	(200 kwh × $0.025 per kwh)
500–600 kwh	2.70	(100 kwh × $0.027 per kwh)
	$17.85	
+ fuel adjustment charge	4.80	(600 kwh × $0.008 per kwh)
	$22.65	

Remember that all the calculations used here are only examples. The rate schedule for the utility that services you is most likely quite different in both price and kilowatt-hour ranges. You will have to contact your electric company to obtain the correct rate schedule for your home. Be sure to specify if you have an electric hot water heater, electric heat, or both. Keep your rate schedule handy by recording the rate schedule for your home in table C-4 (Appendix C).

Natural Gas

Natural gas is sold in 100-cubic-foot (Ccf) units or by the therm. As with the pricing of electricity, the cost per Ccf decreases with the more gas you use per month, unless you live in a state that has lifeline rates. As stated previously, a therm is approximately equivalent to 100 cubic feet, so we will treat them as equivalent. An example of a gas rate schedule (nonlifeline) is shown in table 6.

A fuel adjustment charge of several cents per Ccf may also be added to the rate schedule. For example, a fuel adjustment charge of $0.065 per Ccf would increase the actual costs to those shown in table 7. And, like the fuel adjustment charge on electricity, the

TABLE 6
Example of a Gas Rate Schedule

Range	*Cost*
0–200 cf (0–2 Ccf)	$1.85
Next 1,300 cf (2–15 Ccf)	0.311 (31.1 cents) per Ccf
Next 3,500 cf (15–50 Ccf)	0.243 (24.3 cents) per Ccf
Over 5,000 cf (over 50 Ccf)	0.173 (17.3 cents) per Ccf

fuel adjustment charge for gas is a fixed cost per Ccf and does not depend on your total monthly consumption.

The cost of 6,000 cubic feet of gas per month (60 Ccf) would be calculated as shown in table 8. Because the fuel adjustment charge is usually added to the basic cost of the gas, the rate schedule from table 6 is used and the fuel adjustment charge is added to the basic charge.

An example of a lifeline schedule is shown in table 9, and the calculation of the cost for 120 Ccf is shown in table 10.

Again, all calculations given are just examples. To calculate your own costs, contact your gas company to obtain its rate schedule and fuel adjustment charge (FAC) for your home, and record that information in the appropriate blanks in table C-5 (Appendix C).

Average Costs and Actual Savings

So now you have a head full of numbers—costs per kwh, costs per Ccf, costs for this much and that much, fuel adjustment charges, and so on. What do they all mean, and how can we use them? Because of various rate schedules for electricity and gas, the average cost of electricity and gas can vary with the amount used each month. Therefore when you do manage to save some energy, the calculation of money saved can be rather difficult.

First let's consider fuel oil. It is sold by the gallon and the price

TABLE 7
Example of a Gas Rate Schedule Including the Fuel Adjustment Charge (FAC)

Range	*Cost*
0–200 cf	$1.98 ($1.85 + 2 Ccf × $0.065 per Ccf)
Next 1,300 cf	0.375 per Ccf ($0.311 + $0.065 FAC)
Next 3,500 cf	0.308 per Ccf ($0.243 + $0.065 FAC)
Over 5,000 cf	0.238 per Ccf ($0.173 + $0.065 FAC)

TABLE 8
Sample Calculation of a Monthly Gas Bill for 6000 Cubic Feet

Range	*Cost*
0–200 cf (2 Ccf)	$ 1.85
Next 1,300 cf (13 Ccf)	4.04 (13 Ccf × $0.311 per Ccf)
Next 3,500 cf (35 Ccf)	8.51 (35 Ccf × $0.342 per Ccf)
Next 1,000 cf (10 Ccf)	1.73 (10 Ccf × $0.173 per Ccf)
	$16.13
+ fuel adjustment charge	3.90 (60 Ccf × $0.065 per Ccf)
Total	$20.03

is the same whether you buy 20 or 200 gallons. If you use 1,000 gallons of oil during a year for heat and hot water at 75 cents per gallon, your cost is $750. If you use some conservation measures and reduce your consumption to 900 gallons, you save 100 gallons and $75. You paid for the oil that you used at a rate of 75 cents per gallon, and you saved at the same rate—75 cents per gallon.

But what about electricity and gas? The average cost of 600 kwh in the previous example (table 3) is $0.0456 per kwh ($27.37 ÷ 600 kwh). The costs for 300 kwh and 1,200 kwh per month (half and double of 600 kwh) are $15.25 and $48.36 respectively, using the example rate schedule. Total and average costs for 300, 600, and 1,200 kwh are shown in table 11.

Note that by cutting consumption by 50 percent, you do not get a 50-percent saving; in this example, the saving is about 44 percent. And as you reduce consumption, your average cost goes up.

For gas, the story is similar. Using the example rate schedule in table 7, the total and average costs for 3,000, 6,000, and 12,000 cubic feet are shown in table 12.

TABLE 9
Example of a Lifeline Gas Rate Schedule

Rates:	*Lifeline*	*Nonlifeline*
0–200 cf (0–2 Ccf)	$1.70	$1.80
Next 1,300 cf (2–15 Ccf)	0.202	0.241 per Ccf
Next 6,500 cf (15–80 Ccf)	0.183	0.216 per Ccf
Over 8,000 cf (over 80 Ccf)	0.234	0.272 per Ccf
Fuel adjustment charge (all Ccf)	0.065	0.065 per Ccf

Lifeline quantities:	
Basic use	20 Ccf
Basic use plus heat	100 Ccf (winter), 20 Ccf (summer)

TABLE 10
Sample Calculations for a Monthly Gas Bill for 120 CCF: Lifeline Rates

Basic use only:	
First 2 Ccf	$ 1.70
Next 13 Ccf	2.63 (13 Ccf × $0.202 per Ccf)
Next 5 Ccf[a]	0.92 (5 Ccf × $0.183 per Ccf)
Next 60 Ccf[b]	12.96 (60 Ccf × $0.216 per Ccf)
Next 40 Ccf	10.88 (40 Ccf × $0.272 per Ccf)
	$20.09
+ fuel adjustment charge	7.80 (120 Ccf × $0.065 per Ccf)
Total	$36.89
Basic use plus heat:	
First 2 Ccf	$ 1.70
Next 13 Ccf	2.63 (13 Ccf × $0.202 per Ccf)
Next 65 Ccf	11.90 (65 Ccf × $0.183 per Ccf)
Next 20 Ccf[c]	4.68 (20 Ccf × $0.234 per Ccf)
Next 20 Ccf[d]	5.44 (20 Ccf × $0.272 per Ccf)
	$26.35
+ fuel adjustment charge	7.80 (120 Ccf × $0.065 per Ccf)
Total	$34.15

[a] Last 5 Ccf in basic lifeline quantity (basic use).
[b] Remaining 60 Ccf in "next 65 Ccf" category, nonlifeline rate.
[c] Last 20 Ccf in basic lifeline quantity (basic use plus heat).
[d] Remaining 20 Ccf in "over 80 Ccf" category, nonlifeline rate.

Now how do you calculate the cost to operate an electrical appliance? Simply multiply the kilowatt hours consumed by the cost per kwh for that electricity. But what costs? Are the kilowatt hours consumed from the first 30 kwh of the month, the next 70, the next 400, or the next 500? You cannot specify each kilowatt hour you use as being from a specific category in your rate schedule and thus having some corresponding cost. Instead you must consider the entire month's electrical consumption as one

TABLE 11
Average Costs of Electricity for Different Monthly Uses

Used per month	*Total Cost*	*Average Cost*
300 kwh	$15.26	$0.0509 (5.09 cents) per kwh
600 kwh	27.37	0.0456 (4.56 cents) per kwh
1,200 kwh	48.36	0.0403 (4.03 cents) per kwh

TABLE 12
Average Gas Costs for Different Monthly Uses

Used per month	*Total Cost*	*Average Cost*
3,000 cu. ft. (30 Ccf)	$11.49	$0.0383 (38.3 cents) per Ccf
6,000 cu. ft. (60 Ccf)	20.03	0.334 (33.4 cents) per Ccf
12,000 cu. ft. (120 Ccf)	38.00	0.317 (31.7 cents) per Ccf

large pot full of electricity with some average cost, and any kilowatt hour coming from that pot costs that average amount. Therefore to find the cost of operating an electrical appliance, the kilowatt hours used by the appliance must be multiplied by the *average* cost per kwh. If your refrigerator uses 150 kwh a month and your total electrical consumption is 1,200 kwh per month, the cost of operating that refrigerator is $6.05 per month (150 kwh × $0.0403 per kwh; table 11). But if your total consumption is only 600 kwh per month, that same refrigerator costs $6.84 to operate (150 kwh × $0.0456 per kwh). Therefore the cost of operating an appliance not only depends on the kilowatt hours consumed by that appliance, but also on the total electrical consumption per month. The smaller the total consumption, the higher the average rate and the greater the cost for a given appliance.

Now what about the money saved when you manage to conserve some energy? Unfortunately money savings do not occur at any average rate. The kilowatt hours saved are the last ones you are charged for, and therefore you save at that rate. For example, the cost of 550 kwh and the cost of 600 kwh differ only by the cost of the last 50 kwh (550th to 600th kwh); the first 550 kwh costs the same for both total amounts. From the example rate schedule, including the fuel adjustment charge (table 2), the last 50 kwh costs 3.7 cents per kwh; so if you save 50 kwh, the cost saving is $1.85 (50 kwh × $0.037 per kwh). If your total consumption is between 100 and 500 kwh, the cost saving would be calculated at a rate of 4.2 cents per kwh; between 500 and 1,000 kwh, the saving would be at a rate of 3.1 cents per kwh. These numbers apply only to the example rate schedule (table 2). To find how much *you* would actually save, use the numbers in column D of table C-4 (Appendix C) for the appropriate range of your monthly consumption.

For gas, the same situation exists. You calculate your cost of gas consumed by an appliance at the average cost of the gas; but you calculate the money saved by reducing gas consumption at the cost of the "last" Ccf used.

To repeat: Calculate the cost of operating an appliance at the average rate, but calculate the money saved at the rate for the "last" kilowatt hour, 100 cubic feet, or therm. Although both gas and electricity rates may change slightly from month to month (depending on the fuel adjustment charge), the change will not be sufficient to affect the calculations significantly.

In later examples, we will use an average cost of 6 cents per kwh and 35 cents per Ccf, and a "last" cost of 5 cents per kwh and 30 cents per Ccf. For oil we will use a cost of 75 cents. Although these amounts may be a little higher than the national average (except in some of the large cities), they should represent some approximate average costs over the next few years.

To determine your average rate and the "last" rate for a kilowatt hour or 100 cubic feet, you must resurrect some of your recent utility bills. Your consumption of electricity or gas may vary greatly from summer to winter, so use one bill from a summer month and one from a winter month. Refer to table C-6 (Appendix C) for details on how to find your average and "last" rates for electricity or gas.

Heating Values and Relative Costs of Gas, Oil, and Electricity

We use oil for hot water and for home heat; we use gas for hot water, heat, stoves, ovens, and clothes dryers. We can use electricity to run electrical appliances or to produce heat, but we cannot use oil and gas to run electrical appliances. Therefore if we want to compare the relative costs of the three forms of energy, we can only compare one cost, that of producing heat. Gas is measured in units of 100 cubic feet with 1 Ccf = 1 therm = 100,000 Btu. Oil is measured by the gallon, but recalling that a gallon of oil produces 140,000 Btu of heat, or 1.4 therm, then only 0.71 gallon of oil is needed to produce 1 therm of heat. A kilowatt hour can produce 3,413 Btu of heat; so to produce 1 therm of heat, 29.3 kwh are needed (100,000 Btu ÷ 3,413 Btu per kwh). The costs per therm of heat produced by these three forms of energy are shown in table 13.

Although the numbers in table 13 show the difference in costs for heat produced by oil, gas, and electricity, they can be slightly

TABLE 13
Costs of Energy per Therm of Heat Produced

Fuel		*Cost per Therm*
Oil, per gallon:	$0.75	$0.54
	.70	.50
	.65	.46
	.60	.43
	.55	.39
	.50	.36
	.45	.32
Gas, per Ccf:	.50	.50
	.45	.45
	.40	.40
	.35	.35
	.30	.30
	.25	.25
	.20	.20
Electricity, per kwh:	.06	1.76
	.05	1.46
	.04	1.17
	.03	.88
	.02	.59

misleading. When oil or gas is burned in a furnace or hot-water heater, some of the heating is lost with the exhaust gases—so not all of the heat produced is useful heat. Generally a good furnace or hot-water heater is between 65 and 75 percent efficient—that is, about one-third of the heat is lost, and only two-thirds of the heat ends up as useful in the water or air. When electricity is used for heat, no exhaust gases are produced, and all of the heat produced is useful. Taking these factors into account, the costs per therm of useful heat for home and hot water heating are listed in table 14. The values for oil and gas are calculated for a furnace or hot-water heater with an efficiency of 70 percent, and the values for electricity are based on an efficiency of 100 percent.

The numbers in tables 13 and 14 show comparable rates for oil and gas, but somewhat higher rates for electrical heat. Even at the low rate of 2 cents per kwh, the cost per therm of electrical heat is about the same as the most expensive oil or gas. The reason for this higher cost is that an electrical power plant burns oil, gas, or coal, or uses heat from nuclear fuel to produce electricity. But a power plant is only about 33 percent efficient—only about one-third of the heat energy produced by gas, oil, coal, or nuclear fuel

TABLE 14
Costs of Energy per Therm of Useful Heat Produced

Fuel		*Cost per Therm*
Oil, per gallon:	$0.75	$0.77
	.70	.71
	.65	.66
	.60	.61
	.55	.56
	.50	.51
	.45	.46
Gas, per Ccf	.50	.71
	.45	.64
	.40	.57
	.35	.50
	.30	.43
	.25	.36
	.20	.29
Electricity per kwh:	.06	1.76
	.05	1.46
	.04	1.17
	.03	.88
	.02	.59

is actually converted to electrical energy, while the rest is discarded as "waste" heat to rivers, lakes, oceans, or the atmosphere. Because of the low fraction of original heat value converted to electricity, and because the power companies still must pay for the coal, oil, gas, or nuclear fuel at rates not much lower than a homeowner's cost for gas or oil, the cost of electrical heat will always be higher than that of oil or gas heat in the home.

Efficiency

The efficiency of an appliance is defined as the useful energy divided by the energy input:

$$\textbf{Efficiency} = \frac{\textbf{useful energy}}{\textbf{energy input}}$$

The more useful energy an appliance can deliver for the same energy input, the greater its efficiency. For appliances like a heating system or hot-water heater, the input and output energies are easily defined in terms of Btu, and the efficiency is defined in terms of a fraction or percentage. A furnace with an input energy of 120,000 Btu and an output of 90,000 Btu has an efficiency of 75%:

$$\textbf{Efficiency} = \frac{\textbf{useful energy}}{\textbf{energy input}} = \frac{\textbf{90,000 Btu}}{\textbf{120,000 Btu}} = \textbf{.75}$$

$$= \textbf{75 percent}$$

For some appliances, however, the energy output is hard to define in terms of Btu or kilowatt hour, so the efficiency can also be defined as the job done divided by the energy input:

$$\textbf{Efficiency} = \frac{\textbf{job done}}{\textbf{energy input}}$$

For example, a dehumidifier that uses 1 kwh to remove 2 pints of water from the air has an efficiency of:

$$\textbf{Efficiency} = \frac{\textbf{job done}}{\textbf{energy input}} = \frac{\textbf{2 pints}}{\textbf{1 kwh}} = \textbf{2 pints per kwh}$$

The energy output, or job done, is 2 pints of water removed from the air, and the energy input is the 1 kwh of electrical energy, so the efficiency is not expressed in a fraction or percent but in terms of pints per kwh.

In all cases, the greater the efficiency, the less energy that is needed to do the desired job, whether the job is heating air, heating water, washing or drying clothes, or providing a television picture.

3.

Home Heat

Most of the energy consumed in homes is for heat. The heating system is probably the one appliance that requires the least effort to operate—a flip of the finger turns it on and off—yet it is the most costly to operate. In northern states, home heat can account for about 80 percent of the total energy used in the home. Even in many southern states, annual heating bills can run up to $200 or more. Therefore, a small percentage of home heat saved can mean a lot of money saved over a period of several years.

The amount of fuel required to heat a house depends on many factors—length of the winter season, inside temperature, outside temperature, size of the house, efficiency of the heating system, and ability of the house to retain heat. Size of a house and the severity of winter weather are two things the homeowner can't change without rebuilding the house or moving it to a different area of the country. But the homeowner *can* change the inside temperature and improve the ability of a house to retain heat. Later we'll discuss how to reduce heat loss and consider methods to calculate the amount of heat loss from a house. But before we can proceed to these calculations, we must learn a little about heat transfer and define a few terms.

Heat Transfer

Heat can move in three ways: by conduction, by convection, and by radiation. Conduction is the movement of heat through a solid piece of material. For example, heat transfer through the bottom of a pot on the stove is by conduction. If the pan stays on the stove long enough, the handle also gets hot by conduction through the sides of the pot.

Materials that readily conduct heat are called "good conductors," and those that don't conduct heat well are called "thermal insulators." Most metals are good conductors and are therefore used wherever good heat transfer is desired, such as for the bottom of pots and the boiler in a furnace. But we don't want the pot handle to get hot and burn our hands, so most pots have a plastic or wooden insulating handle. And in our houses we want the heat to stay inside, so we have insulation in the walls and ceilings to reduce the heat conduction to the outside.

A fluid (a liquid or gas) can also conduct heat, but most fluids are relatively poor conductors. But a fluid can move easily, is not restricted to a fixed position as is a solid, and can carry heat with it as it circulates. This movement of heat by movement of a fluid is called "convection."

Most fluids become lighter (less dense) when heated and heavier (more dense) as they cool; thus a heated fluid rises up through the cooler fluid around it. In a baseboard heating unit, for example, as the air in the convector becomes heated, it becomes less dense, rises, and draws more cool air to the heater to be warmed. Because heat moves to the room air by convection, a baseboard heater is really a convector, not a "radiator," as they are commonly called. Simple movement of heat from a convector is an example of "natural convection." If a fan or pump moves the air or water, the convection is "forced convection." Most heating systems use forced convection because the heat flow can be better controlled than with natural convection.

Even though air is a poor conductor of heat, it is a good convector and therefore not a good insulator. For air to become a good insulator, it must be trapped to reduce convection. Fiberglass and other types of insulation work on this principle. The fine strands of fiberglass are themselves poor heat conductors—but more importantly, they trap the air and reduce convection. Fur on animals and the hair on your head trap the air and insulate the body from cold outside air.

The air at the surface of a wall or window forms a thin layer of insulation. During the winter, the air near the inside surface is cooler than the room air, and the air near the outside surface is warmer than the outside air. These thin air layers reduce heat conduction through the window or wall, but if the wind blows away part of the outside layer, this insulating surface looses its effectiveness and heat conduction increases. Therefore, heat loss through a window is greater for a window exposed to the wind than for one sheltered from the wind.

A rather subtle form of convective heat transfer depends upon the evaporation and condensation of water. When water evaporates, it becomes a vapor that moves with the air and condenses elsewhere. Evaporation of a pound of water (about a pint) removes approximately 1,000 Btu of heat from the surrounding material; when the vapor condenses, it releases these 1,000 Btu. Thus heat moves from the area of evaporation to the area of condensation by convection of the water vapor. The energy required to evaporate water is called the "latent heat" of vaporization. We personally experience this evaporative cooling when we perspire on hot days; the evaporating perspiration removes heat from the skin and cools our bodies. On humid days, the evaporation rate is less, so the rate of body heat loss decreases and we feel hotter.

Heat radiation is the movement of heat by the process called "electromagnetic radiation." We know electromagnetic radiation by the more common names of light, ultraviolet waves, radio waves, radar, and microwaves. Whereas heat conduction and convection require some material to transmit the heat, radiation of heat can proceed through a vacuum—much as the light we receive from the sun travels through millions of miles of empty space.

Any object radiates heat, but the form of the radiation depends on the temperature of the object. Objects with temperatures up to several hundred degrees radiate mostly infrared waves. At higher temperatures, part of the radiation is in the form of visible light—the heating element in an electric oven glows red as it heats. At much higher temperatures, a larger portion of radiation is in the form of visible light. The sun, which has a surface temperature of about 10,000° F., radiates most of its energy in the form of visible light. The filament of an incandescent light bulb is hot enough to radiate visible light, while the lower-temperature filament in an infrared lamp radiates mostly infrared waves.

But if every object is radiating heat away, won't everything eventually cool? No, because although all objects in a room, for example, are radiating heat, they are also receiving radiated heat from the other objects in the room. If the temperatures of all the objects are equal then each object receives as much radiated heat as it radiates away—so it has no net heat loss. If one object is warmer than the others, however, then that object soon cools and the others heat up. This is the principle of the common heat radiator and electric radiant heat. Although some heat from the radiator moves to the room by air convection, most of the heat radiates to the room. A fire in the fireplace heats by radiation—the only convection of heat in this case is up the chimney, the warmed air being lost outside. If you put your hand near, but not touching, a cold window, your hand feels cold. But only a small amount of heat is conducted from your hand; most of the heat loss is by radiation. Your hand radiates heat to the cold window, but the cold window doesn't radiate back—so your hand has a net heat loss and feels cold.

The amount of radiation absorbed and radiated by a surface depends on the color of that surface. Any surface with a good heat-absorbing color is also a good radiator of heat. As you probably know, black is the best color for heat absorption, so black is also the best color for heat radiation. White or silvered surfaces are poor absorbers and therefore are also poor radiators of heat.

The fact that objects on earth radiate in the form of infrared waves has important consequences for heat transfer into and out of a house. Visible light passes through glass with very little of it absorbed by the glass; but glass reflects much of the infrared radiation. A glass window allows energy to enter in the form of light from the sun, and objects in the building absorb this light and become warm. These warm objects then radiate infrared waves, but the glass reflects back this radiation inside the building. Therefore the inside of the building becomes warmer because the heat is trapped inside. A car parked in the sunshine gets hot inside—the heat that enters in the form of light can't easily exit because the radiation emitted by surfaces inside the car is in the form of reflected infrared waves.

During the summer when you want to keep your house cool, you should minimize heat transfer into the house by shading your windows from the entry of direct sunlight. During the winter, however, when every little bit of heat in the house helps to maintain warmth, sunlight coming through a window can provide part of your home's total heat needs.

Heating Systems

Most homes in the United States have either a central heating system or electric heating with individual heaters in each room. In a central heating system, the heat produced in a furnace moves throughout the house by convection of either water or air. So before we proceed to heating costs, let's look at the different heating systems and find out how each operates.

Forced Hot-air System

This is the most common type of heating system used today. The furnace consists of a burner (gas or oil) or a heating element (electric) and a fan to circulate air through the house. Figure 4 shows the operation of a forced-air system using either gas or oil for fuel. An electric furnace is similar except that in place of the oil or gas burner, several heating elements provide the heat. And because electric heating produces no exhaust gases in the furnace, an electric furnace needs no chimney.

The circulating fan draws the cold air from the rooms and through the air filter. This cold air flows past the heated walls of the burner enclosure, picks up the heat conducted through these walls, then flows back to the rooms in the house. The exhaust gases from the burned fuel go up the chimney to be vented outside.

The homeowner can adjust heat flow to each room by adjusting the air flow through hot-air registers in each room. Some homes may have zone heating with a thermostat for each zone; different sections of the house can have different temperatures at different times without the need to readjust the hot-air registers each time. Bedrooms, for example, can be kept cool during the day when not in use, while the rest of the house can be kept warm. The basement could likewise be a zone kept cool except when in use.

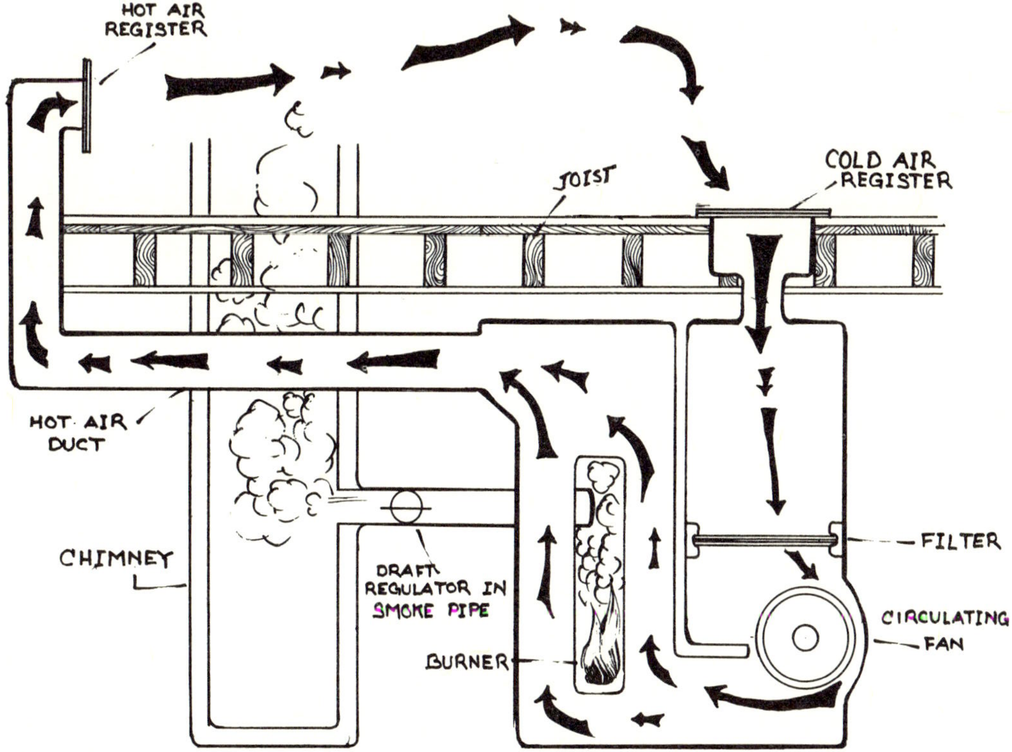

Fig. 4. Forced hot-air heating system (gas or oil)

When one zone requires heat, the furnace supplies it through ducts to that zone only.

One advantage of a forced-air system is its ability to clean the air in the house. The air filter removes the dust and other particles; in addition, some systems have an electrostatic precipitator that removes tiny dust particles, even smoke, from the air. A humidifier installed at the output of the furnace can increase the air humidity in the house. And with an air conditioner installed on the furnace, the entire system becomes a central air conditioner during the summer.

An electric forced-air furnace is 100 percent efficient—all heat produced in the furnace ends up as heat in the house. But if the air ducts pass through an unheated part of the house (attic, crawl space, basement), heat loss can be as high as 20 percent if the ducts are not insulated. A gas or oil furnace is only 65 to 80 percent efficient—20 to 35 percent of the total heat produced in the burner goes up the chimney with exhaust gases and is lost from the house. The efficiency of an oil furnace decreases even more if the burner is not clean; the fuel doesn't burn completely, and some partially or totally unburned fuel goes up the chimney. The filter, if not cleaned or changed regularly, also becomes clogged with dirt and restricts the flow of air. With flow decreased, the circulating air can't pick up all the heat from the burner and some of the heat is lost up the chimney, thus decreasing the efficiency.

Gravity Hot-air System

This is not a common system today; employed with the original coal furnaces, it is usually found only in older homes. The furnace is just like the forced-air furnace except that it has no circulating fan to move the air from furnace to rooms. Instead, the hot air moves by natural convection, so a gravity hot-air system must always be located in the basement. Since the air flow is slow, the furnace must be located in the center of the basement in order to keep the distance minimal to each room. To heat a room far from the furnace, large ducts are required.

A gravity hot-air heating system is less efficient than a forced-air system because the air doesn't move past the burner fast enough for the most efficient heat transfer. Also, any long ducts or bends in the ducts reduce the air flow.

Forced Hot-water System

This is similar to a forced hot-air system except that it uses water to transfer heat to the rooms in the house. Figure 5 shows the operation of a forced hot-water system.

The circulating pump moves the cold water from the room baseboard convectors into the boiler where the water absorbs heat through the boiler walls. The heated water then flows back to the baseboard convectors where the heat transfers to the air in the room. Temperature of the hot water ranges from 140° to 200° F.,

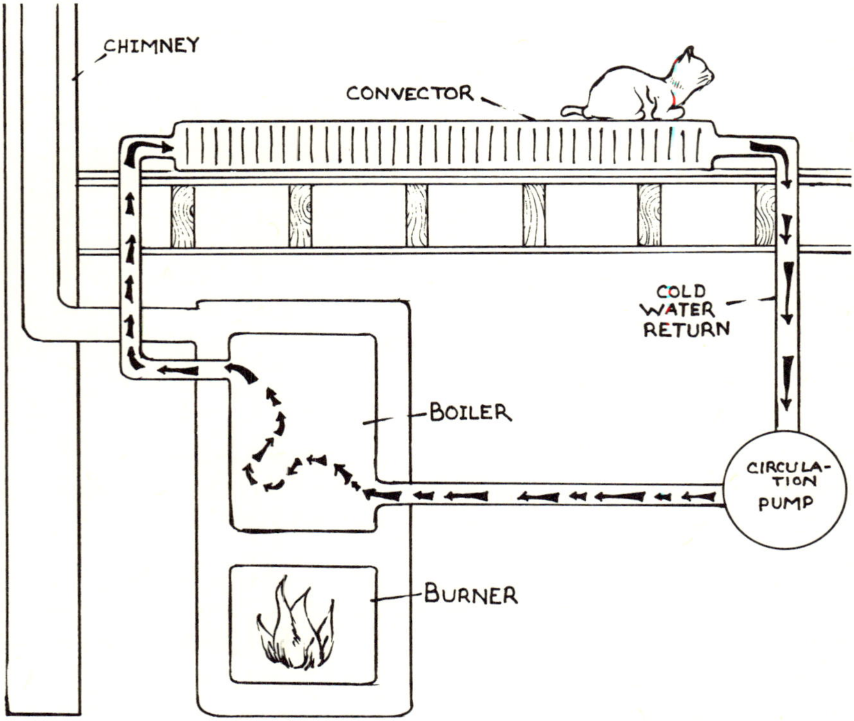

Fig. 5. Forced hot-water heating system

so the water doesn't boil. The efficiency of a forced hot-water system is about 70 percent, although like the forced-air system, that efficiency can drop as the burner becomes dirty.

The heat flow to each room is controlled either by a valve on each convector that controls the hot-water flow to that convector or by an adjustable damper on each convector. If the system is a series loop as shown in figure 6, the only control possible is an adjustable damper. If the system feeds the convectors from a single pipe as shown in figure 7, then either control is possible. One disadvantage of the damper control is that the heat can't be completely turned off in a room.

In a house with zone heating, as previously stated, the different zones have their own thermostats. Heat flow to each zone is controlled either by an independent circulating pump and piping system for each zone or by shut-off valves in the pipes that control the zones to which the hot water flows.

A hot-water furnace can also serve as the hot-water heater for the house. These furnaces have a water pipe that passes through the boiler. The cold water passing through this coil absorbs heat from the boiler water and becomes heated to a temperature of 120° to 140° F. If the boiler is large enough, this type of heater is capable of producing a constant supply of hot water. For these systems, boiler temperature must stand between 160° and 190° F. all year.

One disadvantage of a forced hot-water system is heat loss from the boiler. Because the boiler is at a higher temperature than the surrounding air, the water in the boiler cools when the burner is not operating. This cooling is called the "standby heat loss." If the furnace is in a normally heated section of the house, this heat is not really lost but becomes a source of heat while that section of the house requires heat. But if the furnace is located in an unheated area during times when that area requires no heat, the standby heat loss becomes a real heat loss. And if the furnace also serves as the hot-water heater, this standby heat is also lost during the nonheating season.

Gravity Hot-water System

Like the gravity hot-air system, this system is usually found only in older homes. The system uses two pipes—one supplies

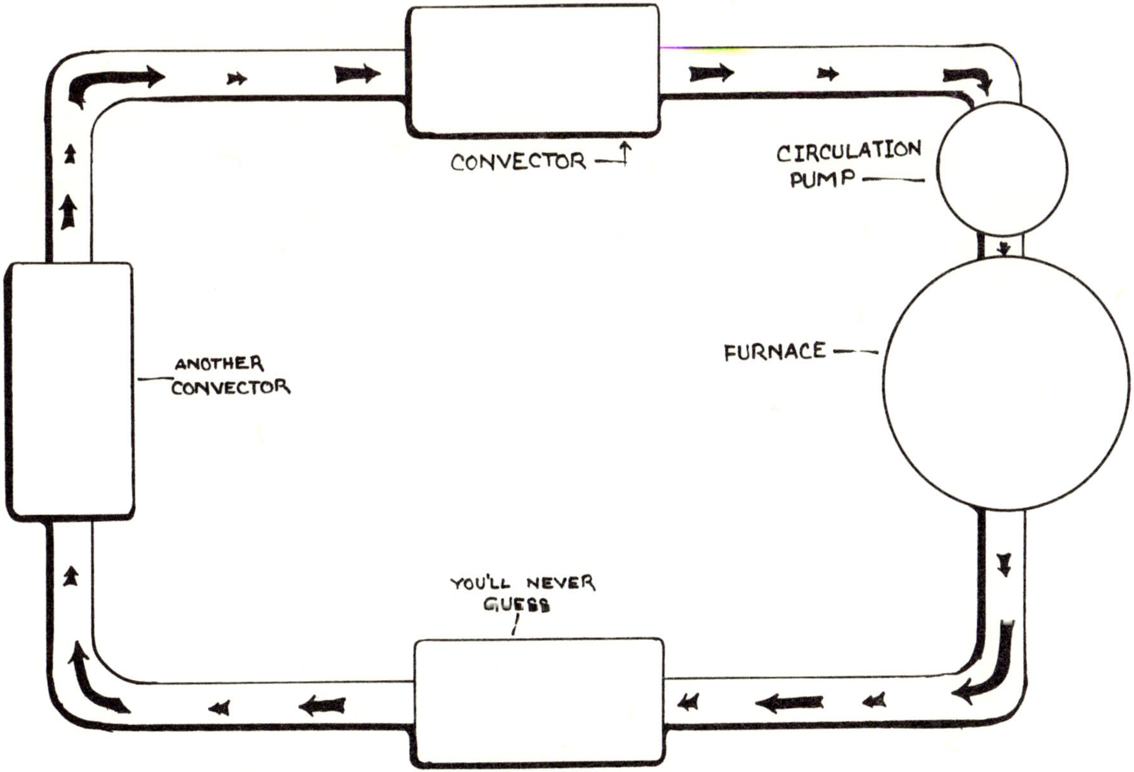

Fig. 6. Hot-water loop system

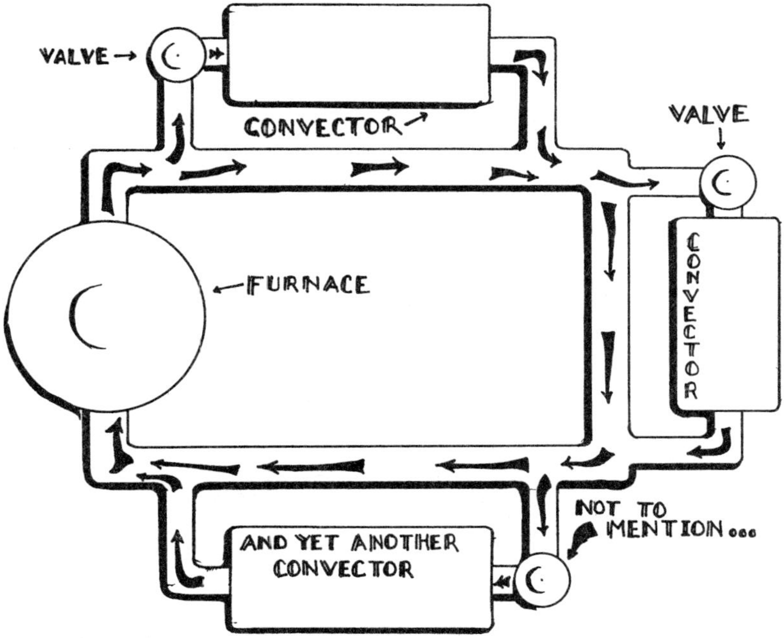

Fig. 7. Hot-water single-pipe system

hot water to each radiator, and the second pipe returns the cooled water to the furnace. The pipes from the furnace to the radiator must be sloped upward to allow the hot water to flow easily by natural convection to the radiators. Any constrictions or too many elbows can restrict water flow and decrease efficiency of the system. A gravity hot-water system is less efficient than a forced hot-water system.

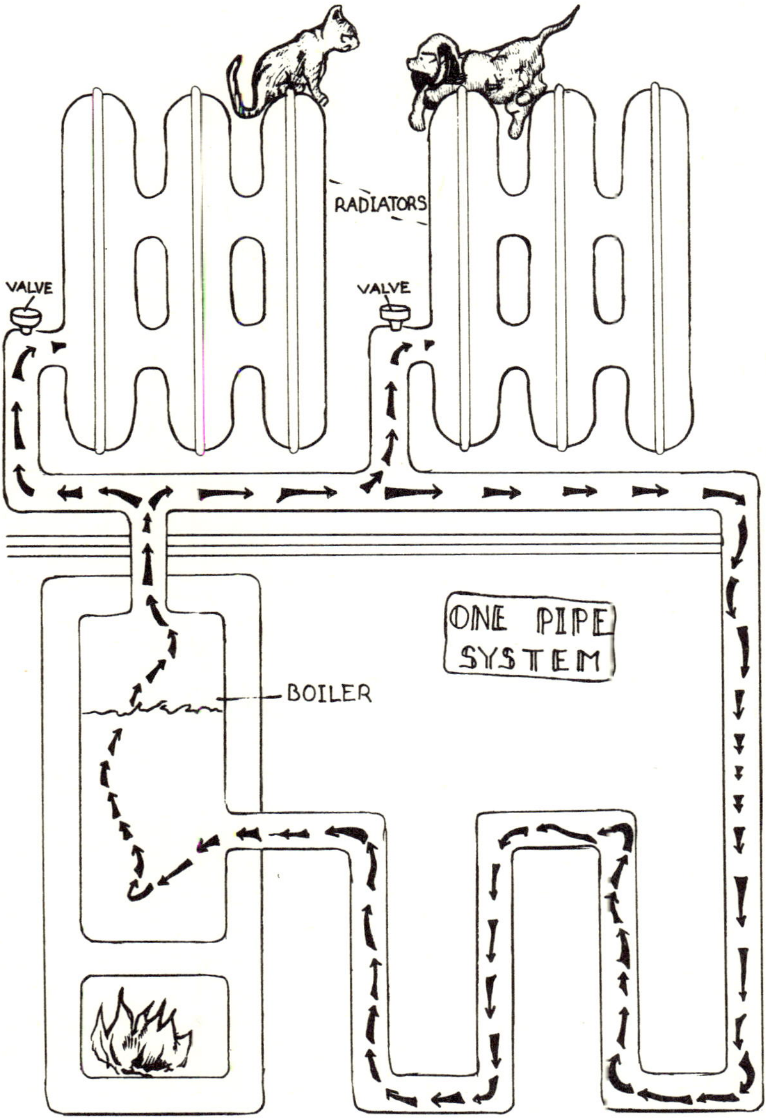

Fig. 8. Steam heating systems

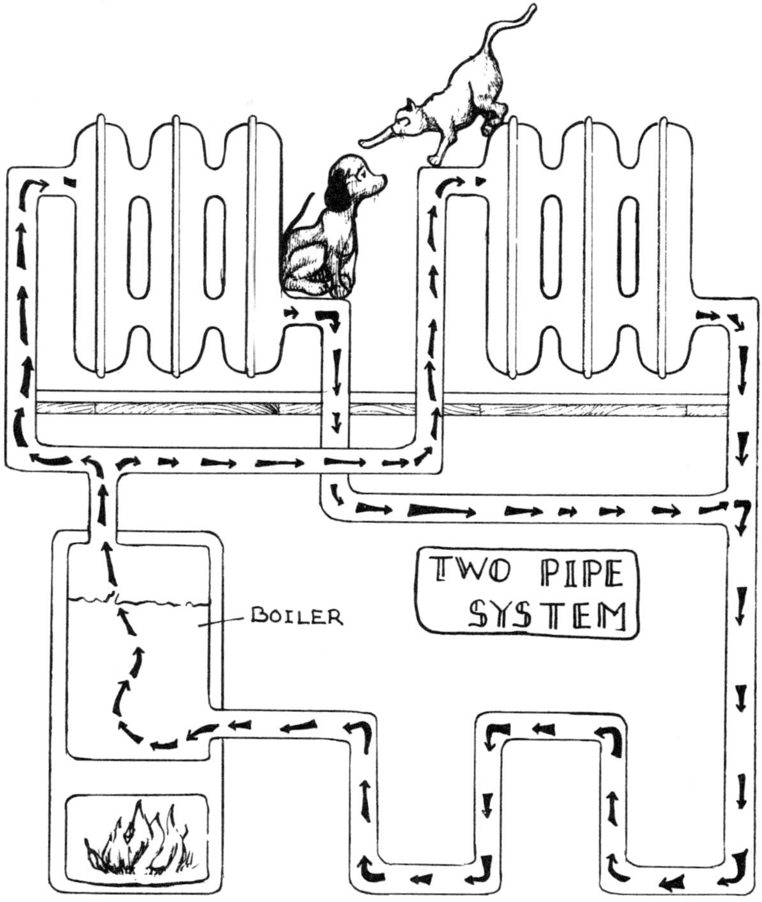

Steam System

This is similar to a gravity hot-water system except that the water boils and steam carries the heat to the radiators. And like the gravity hot-water and hot-air systems, it is fading from the scene because it offers less heat control and costs more to operate than the forced hot-water and hot-air systems.

The operation of a steam system is shown in figures 8a and 8b.

The steam rises to the radiator, looses its heat to the room, and condenses back to water, which returns to the boiler either through the steam pipe in the one-pipe system or through the water-return pipe in the two-pipe system.

The main disadvantage of the steam heating system is its lack of heat control. If the steam is maintained in the boiler, the only heat control is by adjustment of the valves on each radiator. During periods of rapidly changing weather, the valves must be constantly changed to maintain the proper heat control, or each radiator must have its own thermostat to open and close the valve. Because the boiler is at a temperature of 212° F. or higher, the standby heat loss is greater than that of the hot-water system.

Electric Resistance Heat

This is the most popular form of heat for new homes. Electric heat is clean, easy and inexpensive to install, easy to maintain, and permits individual heat control in each room. And before the rapid price rise of electricity, the cost of electric heat was only a

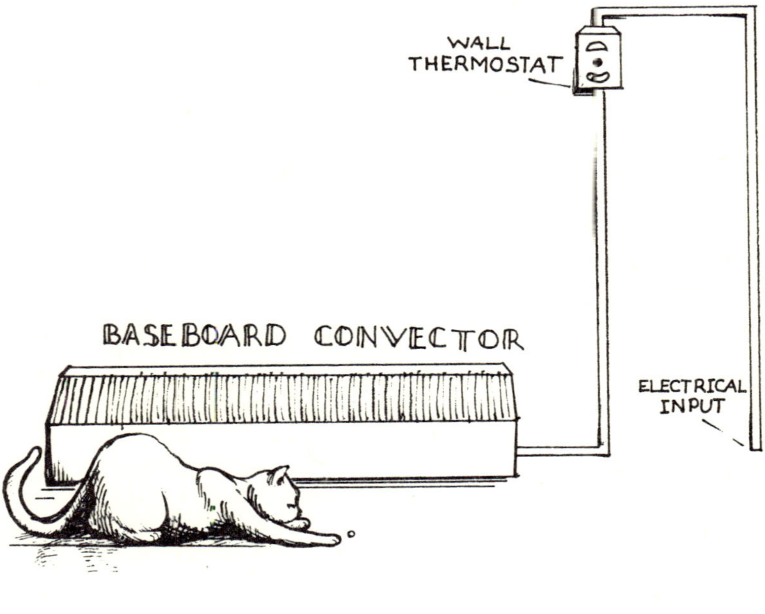

Fig. 9. Electric heating systems

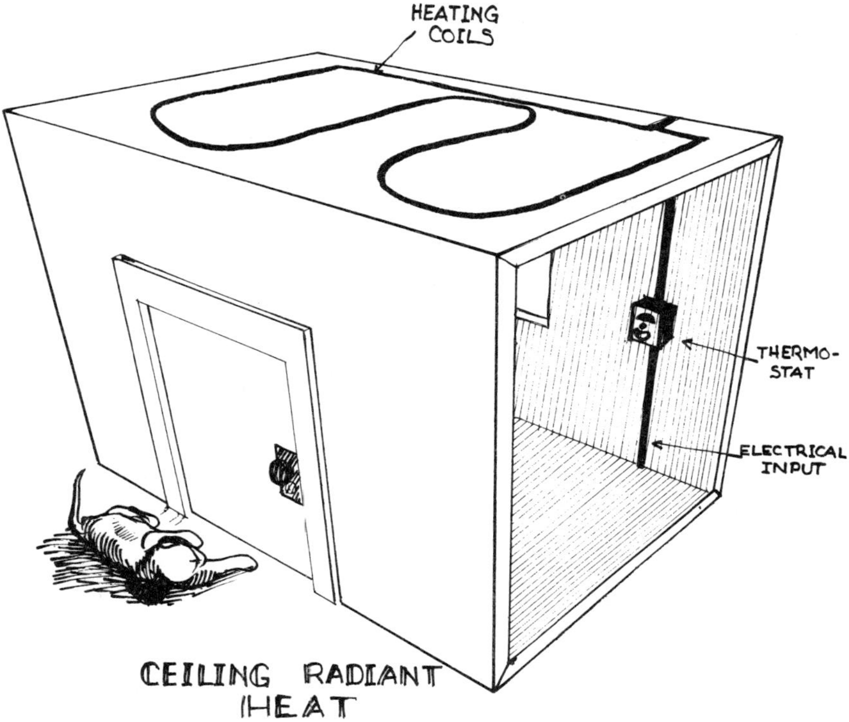

little more than the cost of gas or oil heat. Today, however, homeowners with electric heating may be paying twice what they would pay with oil or gas heating.

Electric heat is 100 percent efficient in the home. All the electrical energy that enters the home is converted to heat—3,413 Btu for each kilowatt hour used. Since the electrical power plant producing the electricity is only about 33 percent efficient, however, electric heat really has total efficiency of only 33 percent. And the heat produced in the home is clean only because the pollution problems of burning coal, oil, or gas have shifted to the power plant.

Electric resistance heating comes in several types: baseboard heaters, wall heaters, and radiant heat. The baseboard heater is similar to a hot-water convector except that the source of heat is an element in the convector rather than hot water from the furnace. A thermostat in each room controls the baseboard heaters in that room. A wall heater, an electric heater mounted on the wall, is just a small heating element with a fan to distribute the heat to the room. Generally wall heaters furnish supplemental heat in bathrooms, entries, or utility rooms and are not the main heating source for a room.

A house with radiant heat has heating cables in the ceiling of each room. These ceiling cables, controlled by a thermostat, also allow individual heating control. When heat is required, an electric current flows through the heating cables. Heat from the cables heats up the ceiling, which then heats the room. The heat transfer to the room is mostly by radiation; air warmed by conduction near the ceiling can't move by convection because it can't rise. To prevent loss of a large fraction of heat through the ceiling by conduction to the attic, the ceiling must be well insulated. As an absolute minimum, a ceiling should be insulated with 9 inches of fiberglass, although 12 inches is preferable. Because the ceiling itself is heated, however, the heat loss will unavoidably be greater than if baseboard heaters provide the room heat.

For the most efficient transfer of heat, the radiating surface should be painted black. Thus black ceilings will provide the best heat transfer to the room in homes with radiant ceiling heat. Yet, because a black ceiling is not aesthetically pleasing and because a light color generally helps keep the room light, a light green or, next best, light yellow is better than white. A green or yellow ceiling radiates at about half the rate of a black ceiling, yet doesn't make the room too dark. A white ceiling radiates only at one-quarter the rate of a black ceiling. You may also use a special ceiling paint for radiant heat ceilings. These paints contain small glass beads that radiate heat better than a normal paint.

Electric Heat Pump

Although not a new type of heating system, the heat pump is just now beginning to enter the heating market. The general definition of a heat pump is a device that takes heat from one place and puts it somewhere else, just like a water pump removes water from a well into a house. An air conditioner is a heat pump that moves heat from the inside of the house to the outside; the refrigeration unit in a refrigerator is also a heat pump, transferring heat from inside the refrigerator to the room air. Similarly, a heat pump used for home heating removes heat from outside and sends it inside—like a reversed air conditioner.

But isn't winter air cold? Is there any heat at all out there? Yes, there is—a surprising amount of it. Until the temperature drops to −460° F. (the lowest temperature possible), all air contains heat, and the heat pump can move that heat inside. Think about your freezer with its inside temperature of 0° F. When the freezer is operating, the outside coils are warm; that warmth is the heat removed from the air inside the freezer. You can think of a heat pump as a freezer turned inside out.

Operation of a heat pump is really quite simple. The unit consists of a few basic parts and a special fluid, usually freon. The mechanism, shown in figure 10, consists of a small valve, two sets of cooling coils, two fans, and a compressor driven by a motor. The operation is based on a simple law of gases—when a gas is compressed, it heats up. (Have you ever touched a tire pump after compressing the air to pump up a tire?) And conversely: when a gas expands, it cools. (Touch the top of an aerosol can after you've used it for awhile.)

The compressor compresses the freon, and the temperature increases to about 130° F., which is higher than the inside air temperature. Inside air blowing across the hot cooling coils removes some heat, and the freon cools to a temperature of about 80° F. As the freon passes through the expansion valve (a small hole that allows only a little freon to pass through), the fluid expands and cools to a temperature below the outside temperature, about 0° to −40° F. (depending on the particular design of the heat pump). When the outside air blows across the outside cold cooling coils, the freon (now colder than the outside air) picks up heat from the air and warms to a temperature of 5° to 10° F. below the outside air temperature. The compressor again heats the freon, the temperature again rises to 130° F., and the cycle repeats.

The heat from the heat pump has two sources—outside air and the electrical energy used to run the compressor. Because the heat from the outside air comes from "nowhere," that is, not

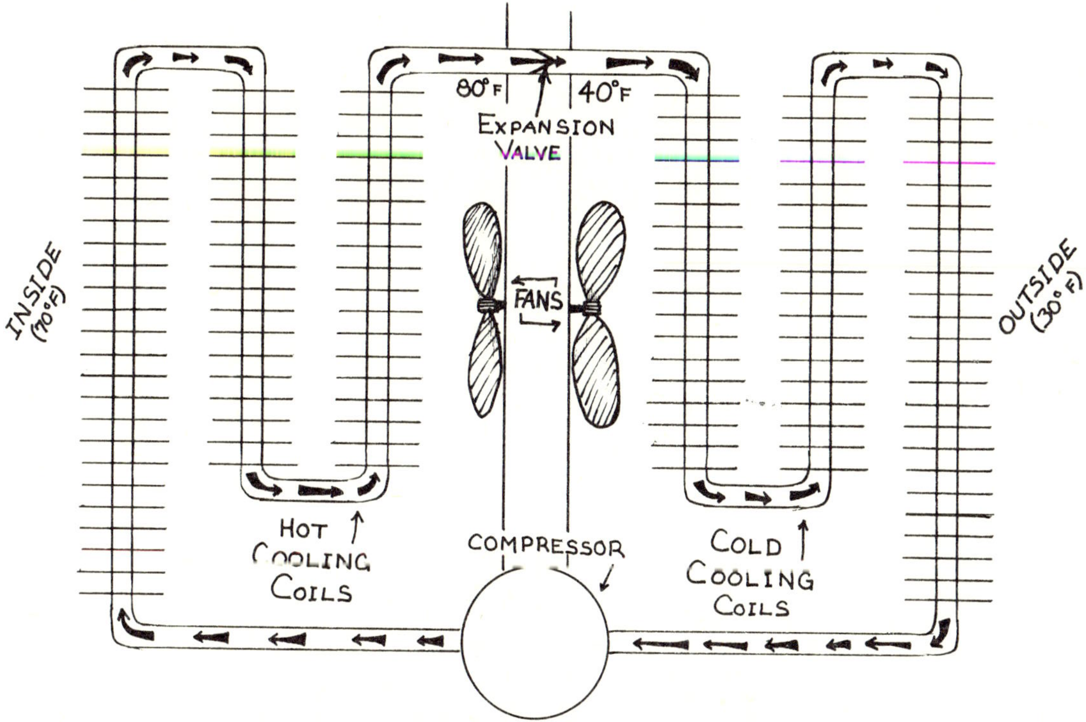

Fig. 10. Operation of a heat pump

from a fuel we can measure and pay for, the efficiency of a heat pump is hard to define. The useful energy is the heat that comes into the house, and this energy is greater than the electrical energy used to run the heat pump. If we define "efficiency" as the useful energy divided by the energy input, the heat pump's efficiency is greater than 100 percent. But the heat pump is not converting more than 100 percent of the electrical energy to heat. While all of the electrical energy used to run the compressor is converted to heat, in the process the heat pump brings extra heat into the house from the outside; thus the useful energy will always be greater than the input energy. Scientists are peculiar in disliking anything to have an efficiency greater than 100 percent, so they redefine terms and call the "efficiency" the coefficient of performance (COP).

The COP of a heat pump depends on the relative temperatures of inside and outside air. At normal room temperatures of about 70° F. and for an outside temperature of about 30° F., the COP is about 2—for each unit of electrical energy used, two units of heat enter the house—one from the electrical energy to run the compressor and the second unit from the outside air. With a COP of 2, in other words, you get twice as much heat from a given amount of electrical energy than you would from electrical resistance heating. The COP increases almost to 3 at an outside temperature of 70° F., but for outside temperatures below 30° F., the COP decreases; and at an outside temperature of 0° F., the COP is nearly 1. The COP never decreases below 1; that is, the electrical energy input will always provide its heat, so a heat pump will never cost more to operate than electrical resistance heating. Most heat pumps have make-up heaters for the very cold days when the pump can't supply the needed heat. Except for regions where outside temperatures frequently reach subzero values more than 85 percent of the time, however, the heat pump can supply the needed heat.

One advantage of a heat pump is that during summer it becomes an air conditioner. A valve installed in the freon pipes changes the fluid direction so that the cold cooling coil is inside the house and the hot cooling coil is outside, as shown in figure 11.

One disadvantage of a heat pump is the relatively low air temperature of the heated air. People are used to air temperatures from forced-air furnaces of 120° to 150° F. and like to feel nice and

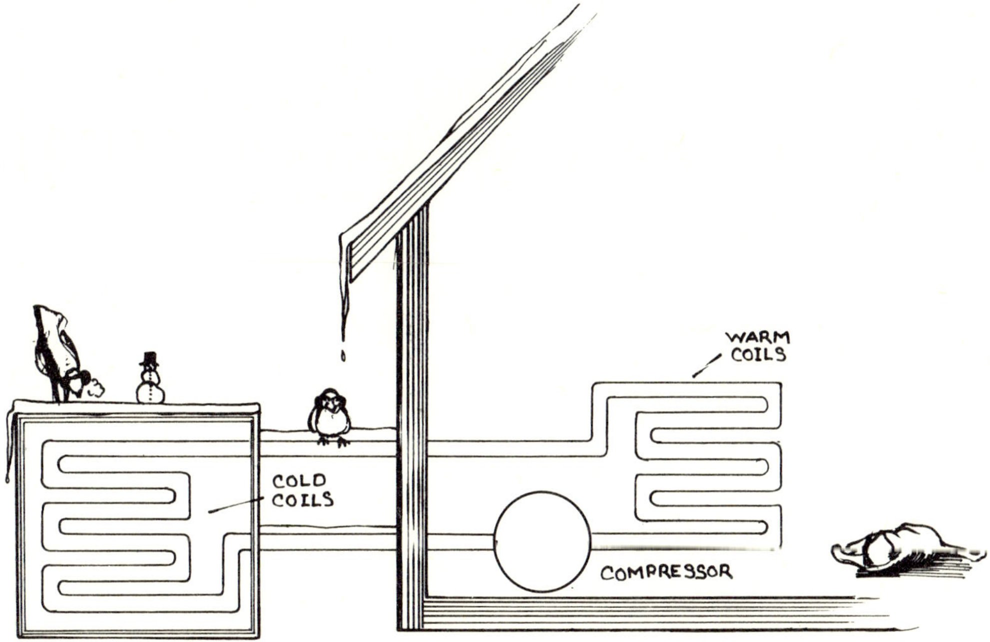

Fig. 11. Heat pump...

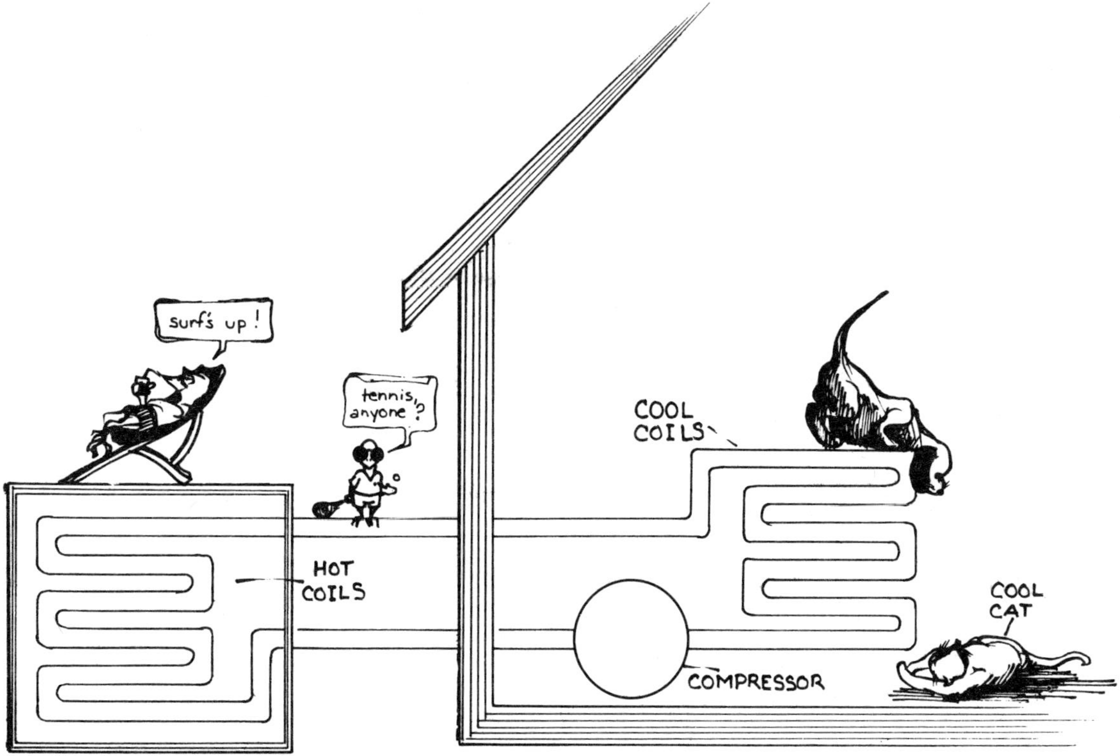

... converted to an air conditioner

warm when they pass in front of a hot-air register. But temperature of the warm air from a heat pump is between 85° and 90° F. when the outside temperature is 30° F. When the outside temperature is only 10° F., the air from the heat pump is only about 80° F. Thus homeowners with a heat pump are tempted to turn on an auxiliary electric resistance heater to increase the air temperature; but this action reduces the total efficiency of the system.

To see how efficiency can decrease with the use of auxiliary heaters, consider a heat pump that can produce 15,000 Btu of heat per hour when the outside temperature is 25° F. At this temperature, the COP is 1.8 and the unit uses 2,440 watts. The air from the heat pump is at a temperature of 85° F., but with a 5,000-watt auxiliary heater (producing an additional 17,065 Btu per hour), the air temperature rises to 105° F. More comfortable, perhaps, but the unit now uses 7,440 watts and the total COP drops to 1.26.

Solar Heating Systems

Until a few years ago, only a few homes had solar heating systems. Since solar energy is the only form of energy that hasn't increased in price, however, these systems are becoming more popular. The operation of a solar heating system is the same as a forced hot-air or hot-water system with a solar collector substituted for the furnace. Figure 12 shows a basic water and air system.

Unfortunately the sun has a habit of not shining at night and hiding behind clouds for much of the time; so a heat storage system is required to provide heat on cloudy days and at night. For a solar hot-air system, the usual storage unit is a large pile of rocks. The heated air from the solar collector passes through the rocks and heats them to 150° F. or more. These rocks retain their heat; on cloudy days and at night, the room air circulates through these heated rocks and picks up the heat. For solar hot-water systems, the storage unit consists of a large tank of water. Figure 13 shows a typical water and air solar system with heat storage units.

The amount of heat that a solar heating system can deliver depends on the particular design of the collectors, their size, and the average number of sunny days during winter. In the southern United States, a relatively small collector area could furnish all the heating requirements for a house; but in the far north,

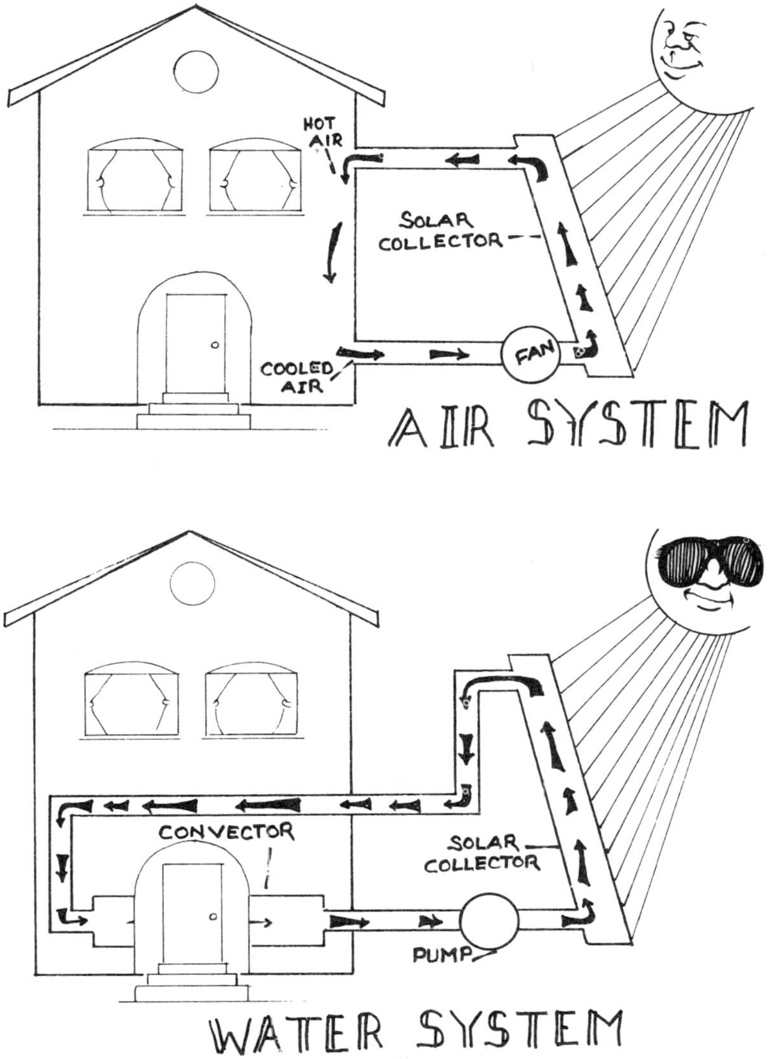

Fig. 12. Solar heating system—air and water

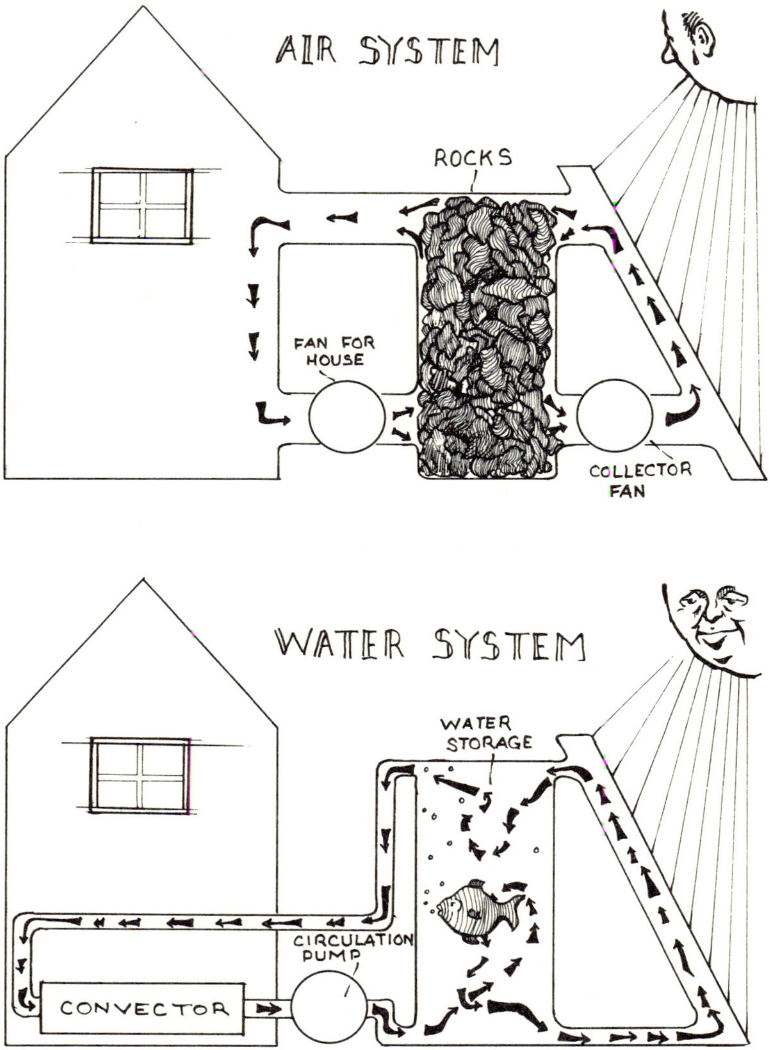

Fig. 13. Solar heating systems with heat storage

even a large collector could hardly supply all home heating requirements.

Disadvantages of a solar heating system include the large initial cost of the system and bulkiness of the materials. The solar collectors are large, expensive, and would probably cover the entire south side of a house roof. A water system needs antifreeze in the collector water or a system to drain the collector water so the pipes don't freeze at night. The system could have corrosion problems. The heat storage part of the system is also large and expensive. And if the solar heating system can't supply all the required heat, a conventional heating system is needed as a backup. In regions where other fuel costs are high, however, the large investment in solar heating can be paid off in a few years with savings on conventional fuel. Many companies are beginning to produce different styles and variations of solar heating systems. If you are considering a solar heating system, you should learn as much about solar heating as possible, then investigate the different designs to see which would be best for your home.

One important aspect of solar heating is house design. The house should have as much insulation as possible; it should be properly oriented with respect to sun and wind directions; and the windows should be properly sized and located to maximize heat gain and minimize heat loss. Some houses even incorporate a large stone wall inside to absorb heat from the sun and act as a heat storage system. The number of ideas and concepts associated with solar heating is limited only by the originality of persons interested in solar heating. If you want to know more about solar heating, the best way to learn is to read several books on the subject.

Efficiency of Your Heating System

To calculate the energy needed to replace the heat loss from your home, you must know the efficiency of your heating system. You may skip this section if you have electric heat, since that system is 100 percent efficient. But if you have oil or gas heat, continue reading.

To find your heating system efficiency, you need two numbers from your furnace instructions—the gross input (Btu input per hour,[1] cubic feet of gas per hour, or gallons of oil per hour) and the net Btu per hour output or Btu at bonnet. The efficiency, as previously defined, is the output divided by the input; so in this case, efficiency is the Btu per hour output divided by the Btu per hour input. Instead of efficiency, however, we really need to find the useful heat per gallon or per Ccf to calculate the gallons of oil or Ccf of gas required to replace heat losses from the house. To find the useful heat per gallon or per Ccf, just multiply the efficiency by the heat produced from one gallon of oil or one Ccf of gas. A furnace with an output of 120,000 Btu and an input of 170,000 Btu per hour has an efficiency of 71 percent (120,000 ÷ 170,000). If the fuel for the furnace is oil, the useful heat per gallon is 99,000 Btu per gallon (0.71 × 140,000 Btu per gallon); and if the fuel is gas, the useful heat per Ccf is 71,000 Btu per Ccf (0.71 × 100,000 Btu per Ccf).

Look up the input and output values for your furnace, record them in table C-7 (Appendix C), and calculate the efficiency and useful heat per gallon of oil or per Ccf of gas. If you have any questions about the numbers or can't find them, consult your heating contractor or service representative.

Your Energy Use for Heating

To determine your heating costs for a heating season, you must know the amount of energy used to heat your house. If you have oil heat and don't use oil for anything else, your total oil consumption for the year is obviously only the oil you use for heat. But if you have oil heat and an oil hot-water heater, gas heat and other gas appliances, or electric heat, you must separately calculate the energy used for heat from the energy consumed for other appliance usages.

We can generally assume that the amount of energy used by all other appliances, except air conditioners and other seasonal appliances, is nearly constant throughout the year; the amount of energy used by these appliances during the nonheating season should closely approximate the energy used by them during the heating season also. If you have oil or gas heat, you can consider the summer months' fuel consumption as representative of the oil or gas amounts used by the other appliances. Likewise, if you have electric heat but no air conditioning or dehumidifiers, the summer months' bills will indicate the amount of year-round nonheating electric usage. With electric heat plus an air conditioner or dehumidifier used only during the summer, however, you can use for calculation only those months when neither heat, air conditioner, nor dehumidifier is used, or used only slightly. Use table C-8 (Appendix C) to calculate the cost of the energy you use for home heat.

Heat Loss; Conductivity, Convection, and Degree-Days

Now that you know how to get heat into your house and how much energy you use to heat it, let's see how to calculate the amount of heat that leaves the house through windows, doors, walls, floors, and ceiling. Although major heat loss is by conduction through these surfaces, a house with leaky window and door seals can also lose much heat by convection, especially if the house is located in a windy area.

Heat loss by conduction through a surface, as measured in Btu per hour per square foot, depends only on the conductivity of that surface (that is, the ability of that surface to conduct heat) and on the temperature difference between inside and outside air:

$$\textbf{Btu per hour per square foot} = \textbf{Temperature difference} \times \textbf{conductivity}$$

The conductivity depends on wall thickness and on the ability of wall materials to conduct heat. Instead of the conductivity of a material, however, we generally specify the resistance to heat flow, or R-value. The R-value is the inverse of the conductivity (R = 1 ÷ conductivity), so a higher R-value means a lower conductivity. With the R-value, the equation for heat loss per hour per square foot becomes:

$$\textbf{Btu per hour per square foot} = \frac{\textbf{Temperature difference}}{\textbf{R-value}}$$

The R-value of a wall consisting of two or more layers of different materials is a combination of the R-values of the different materials. For example, consider a typical wall in a house as shown in figure 14. The R-value (R = 14.19) is for the section of the wall with insulation. Because the R-value of the studs (R = 4.35) is less than that of the insulation (R = 11.0), conductivity through the studs is greater than through the insulated area. The R-value for the total wall is a combination of the R-values for the insulated area (7/8 of the total wall area) and for the studded area

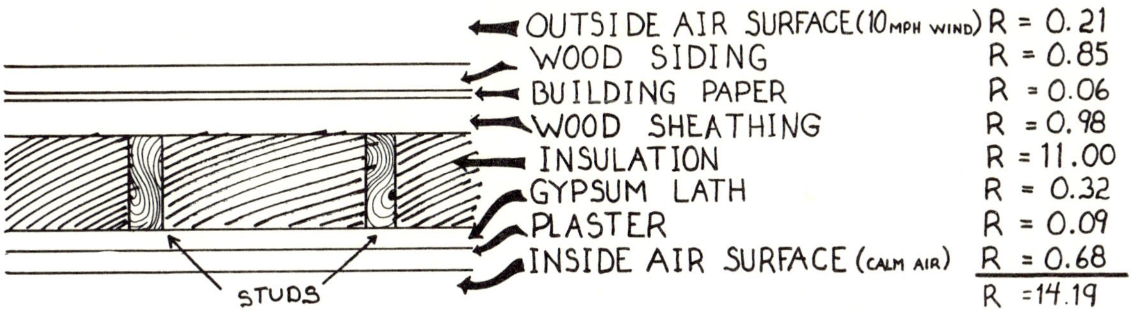

Fig. 14. R-value for a wood-siding wall

(1/8 of the total wall area); thus the R-value for the entire wall is 12.8.[2]

Insulation for walls and ceilings comes in a variety of forms—vermiculite, perlite, fiberglass, mineral wool, cellulose fibers, rigid plastics (polystyrene and polyurethane), and urea-formaldehyde foam. The basic function of each form of insulation is to trap air, thereby reducing the convection of heat. While vermiculite, perlite, fiberglass, mineral wool, and cellulose fibers do trap the air, they cannot prevent a little air movement through the insulation. Rigid plastic and foam insulations, however, contain small air bubbles that cannot move at all; thus these materials are better insulators than the other forms. R-values of these materials are given in table 15.

The common thicknesses of blankets and batts of fiberglass insulation are 3½ inches (R-11) and 6 inches (R-19). For better insulation, the 3½-inch and 6-inch thicknesses can be combined to give 7 inches (R-22), 9½ inches (R-30), 12 inches (R-38), and so on. The loose-fill insulation can be installed to any desired thickness and can be blown into an uninsulated finished wall space. Although the best insulations are polystyrene and polyurethane,

TABLE 15
R-value of Different Insulations

Type of Insulation	*R-value per inch*[a]
Loose fill:	
Vermiculite	2.3
Perlite	2.9
Cellulose fiber	3.8
Blankets or Batts:	
Fiberglass	3.3
Mineral wool	3.3
Rigid Plastic:	
Polystyrene:	
Molded beads	3.9
Extruded: 1.8 lb./cf	4.2
Extruded: 3.5 lb./cf	5.6
Polyurethane	6.3
Foam:	
Urea-formaldehyde	5.3

[a]All values in this table are for a temperature of 50° F. The R-value increases for lower temperatures and decreases at higher temperatures. Data from American Society of Heating, Refrigerating and Air Conditioning Engineers (ASHRAE), *Handbook of Fundamentals* 1972.

these must be carefully used and installed according to manufacturers' specifications.

The R-value of any surface includes the R-value for the surface layer of air. For a wall, ceiling, or floor, the R-value for the surface air is only a small part of the total R-value. But since glass has a low R-value (R = 0.03 for 1/8-inch glass), the glass itself offers little insulation, and the surface layer of air is the major part of a window's R-value. A single pane of glass in a 5-mph wind has an R-value of 1.01 (0.30 for the outside surface + 0.03 for the glass, + 0.68 for the inside air surface). In a 20-mph wind, the R-value for the outside air surface is only 0.14, so total R-value for the window is 0.85. For a storm window, the trapped air between window and storm window has an R-value of 1.0, so the R-value for window with storm window is 2.01 in a 5-mph wind and 1.85 in a 20-mph wind. Note that a storm window reduces the heat loss from a window by about 50 percent. But a window area, even with a storm window (R = 2), loses much more heat per area than a wall (R = 13).

Heat loss by convection (infiltration of air) is not as easy to calculate as heat loss by conduction. Air leaks out of the house through cracks around windows and doors and is replaced by outside air that must be heated to the inside temperature. For each cubic foot of air that infiltrates the house, 0.018 Btu are needed to heat that air by 1° F. When the outside air is 30° F. and the inside air is 70° F., 0.72 Btu are needed to heat each cubic foot of incoming air.

But how much air infiltrates the house? The answer to this question is impossible to calculate exactly. The amount of infiltrating air depends on the wind velocity, number of doors and windows, and size of the cracks around each window and door. Because such cracks are small and difficult to measure, you can't easily calculate the volume of air passing through them. As an indication of the volume of air that infiltrates a house, a well-sealed house may have 20 percent of its air replaced every hour on a calm day, while over 100 percent of the air may change per hour in a poorly sealed house. On windy days these air change rates are higher. A house with 1,500 square feet of floor area, an 8-foot ceiling (12,000 cubic feet of air), and an air change rate of 100 percent per hour has a heat loss of nearly 9,000 Btu per hour when the outside temperature is 30° F.

Heat loss from both conduction and convection depend on the temperature difference between inside and outside air. Most homes have a constant inside temperature; or if temperature is reduced at night, at least the average daily temperature is constant throughout the winter. Outside temperatures vary considerably, however, both daily and seasonally. Figure 15 shows the variation in outside temperature for a day during winter. Although heat loss at any specific time is proportional to the temperature difference at that time, the heat loss for a day is proportional to the difference between average inside and average outside temperatures.

The National Weather Service records and reports the average daily temperature at each reporting station; you can usually find these values in your local newspaper or on the local nightly television news and weather broadcast. In addition to the average daily temperature, the Weather Service reports the degree-days for the day. A degree-day is the temperature difference between

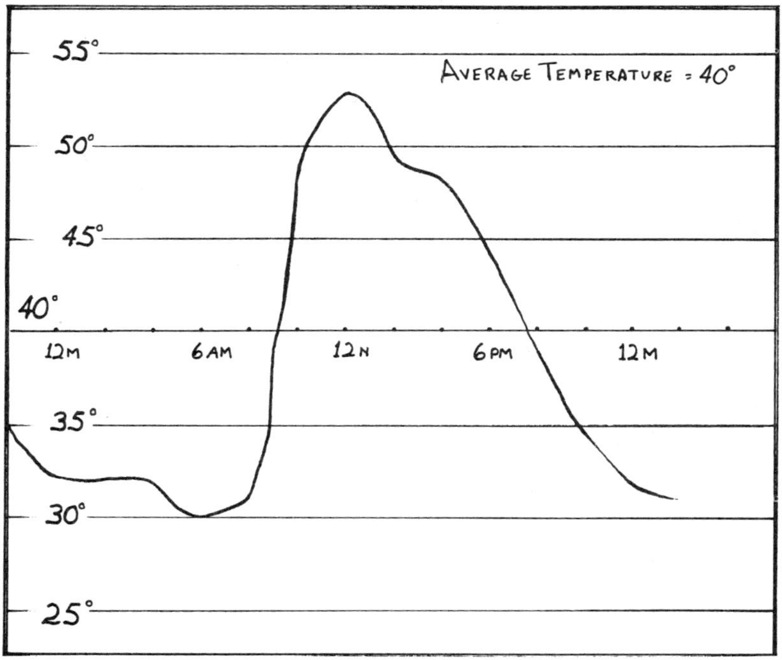

Fig. 15. Outside temperature variations during a day

the average outside temperature and 65° F. The average temperature for the day shown in figure 15 is 40° F., so the degree-day total for that day is 25 (65° − 40° F.).

The degree-day is proportional to the amount of fuel needed to keep a house heated. Conductive and convective heat losses are proportional to the difference between the average outside temperature and the average inside temperature (assumed to be 70° F. for an average home); so why is the degree-day calculated at an average inside temperature of 65° F. rather than 70° F.? The reason is that the degree-day is based on the amount of fuel needed to keep a house heated—and although the home heating system provides most of the heat for the home, it doesn't provide all of it. Heat from electrical appliances used in the home, heat from people in the house, and heat from sunlight entering through the windows also provide a significant amount of home heat. The heat from these sources is enough to make up the temperature difference between 65° and 70° F.

The total number of degree-days for a heating season vary greatly over the country—from a few hundred in the far south to over 10,000 in the far north. Annual degree-days for several major American cities are: New York, 5,000; Atlanta, 3,095; Chicago, 6,300; Minneapolis, 8,310; Dallas, 2,290; Denver, 6,016; San Francisco, 3,042; and Seattle, 5,185. A house in Denver uses about twice as much energy for heat as the same house would use in Atlanta. Figure 16 shows the average number of degree-days for the continental United States.

Heat Losses from Your House

Now we're ready to find the amount of heat that is exiting through your own walls, windows, ceilings, and floors. To do the calculations, you must know the average outside temperature during the heating season; the average inside temperature; the length of the heating season; the R-value and area of each wall, ceiling, floor, and window. Once we know all these quantities, we can calculate the total conductive heat loss for the year through each surface with the following equation:

$$\textbf{Heat loss per heating season} = \frac{\textbf{Area} \times \textbf{hours in a heating season} \times \textbf{average temperature difference}}{\textbf{R-value}}$$

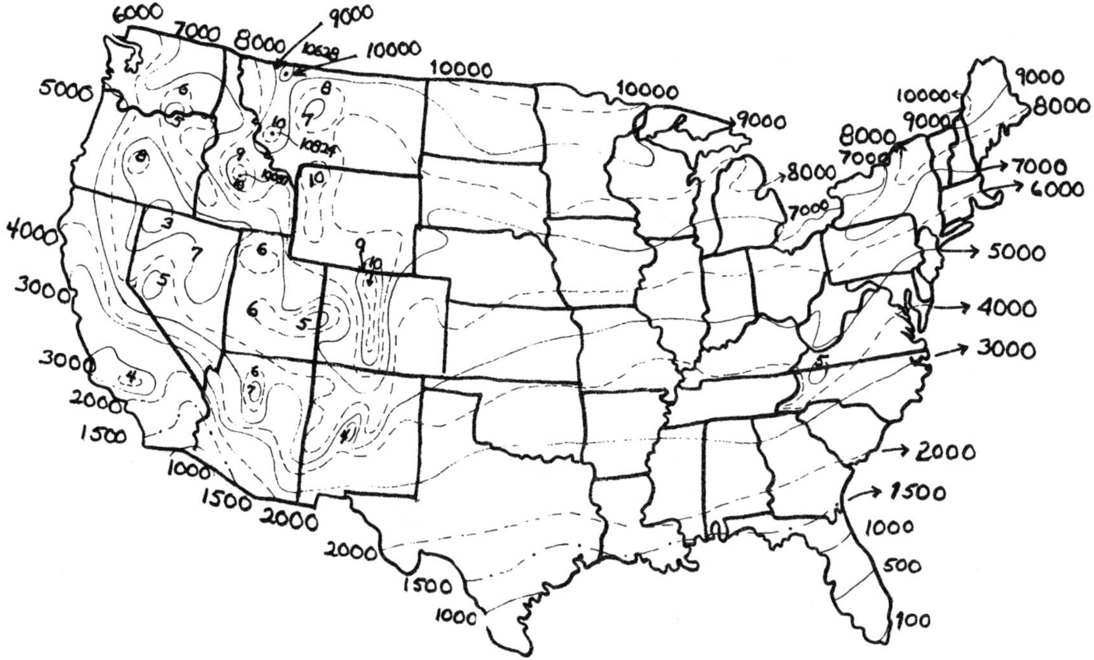

Fig. 16. Annual degree-days for the United States

Let's start inside where it's warm. If you keep the temperature constant day and night, then that temperature is the daily average inside temperature; record that value in the last blank of table C-9 (Appendix C). But if you decrease the temperature at night and possibly set the temperature at some different value during another part of the day, fill in the appropriate blanks of table C-9 and calculate the average inside temperature.

Now let's step outside and determine the average outside temperature for the heating season—and at the same time determine the length of the heating season. Table D-1 (Appendix D) lists the degree-days, number of heating days, and average outside temperature of the heating season for many cities in the continental United States and Alaska. Record the degree-days, number of heating days, and average outside temperature for your locality in table C-10 (Appendix C); then calculate the Degree-difference-hours for your house.

The Degree-difference-hours (Ddh) figure is the total number of hours in the heating season multiplied by the temperature difference between average inside and average outside temperatures. For example, a house that has an average inside temperature of 68° F. and is located in Chicago would have a heating season of 247 days and an average outside temperature of 40° F. (from table D-1, Appendix D). The difference between the inside and outside temperature is 28° F. (68° − 40° F.), so the Degree-difference-hours for this house would be 28° F. × 247 days × 24 hours per day = 166,000 Ddh. In later examples of heat loss calculations, we will use this value for the Ddh. Other cities with approximately the same Degree-difference-hours (for an average inside temperature of 68° F.) include Colorado Springs, Colorado; Hartford, Connecticut; Detroit, Michigan; Omaha, Nebraska; Wilkes-Barre, Pennsylvania; Milford, Utah; and Redmond, Oregon.

Substituting the Degree-difference-hours for the product of hours in a heating season and average temperature difference in our heat loss equation, the final form of the heat loss equation looks like this:

$$\textbf{Heat loss per heating season} = \frac{\textbf{Area} \times \textbf{Ddh}}{\textbf{R-value}}$$

The only information we now need to calculate heat losses is the R-value and the areas of walls, windows, ceilings, and floors.[3] To find the areas, you will have to do some measuring, so use your trusty ruler or tape measure to space off the dimensions of your house. Sketch the outside of your house on a separate piece of paper, and indicate its outside dimensions and dimensions of windows and doors on your sketch. All measurements should be in feet, so convert all inches to fractions of feet.[4] The R-values are given in the appropriate sections that follow.

Windows—Conduction

R-values for the different types of windows are given in table 16. These values are for a 10-mph wind. Note that the values are different for wood and for aluminum frames. A single-glazed window is a window with only one pane of glass and no storm window. A double-glazed window (sometimes called a thermopane) has two panes of glass sealed together with a 1/4-inch or 1/2-inch gap of air between them.

Remember that the amount of heat loss through a surface is the area of the surface multiplied by the Ddh for the house and divided by the R-value; so for a 10-square-foot (ft^2) single-pane aluminum window, the annual heat loss would be 10 ft^2 × 166,000 Ddh ÷ 0.9 (R-value) = 1,840,000 Btu (for our example house in Chicago). If the house has oil heat and the furnace produces 99,000 Btu per gallon of useful heat, the oil needed to replace the heat loss through this window would be 19 gallons (1,840,000 Btu ÷ 99,000 Btu per gallon), which would cost $14. A gas furnace that produces 71,000 Btu per Ccf of useful heat would use 26 Ccf (1,840,000 Btu ÷ 71,000 Btu per Ccf) at a cost of $9. If the house

TABLE 16
R-values for Windows

Glazing	*Type of Frame*	
	Wood	*Aluminum*
Single	1.0	0.9
Single with storm	2.0	1.9
Double (¼-inch gap)	1.7	1.5
Double (½-inch gap)	1.9	1.7
Double with storm (¼-inch gap)	2.7	2.5
Double with storm (½-inch gap)	2.9	2.7

had electric heat, the heat loss would be 540 kwh (1,840,000 Btu ÷ 3413 Btu per kwh) at a cost of $32.

If you have a sliding glass door, consider it as a window. All sliding glass doors have double-glazed glass, but the R-value of even double-glazed glass is much less than the R-value for a wall, so heat loss can be considerable. A 7- × 5-foot glass door would have an annual heat loss of 3.4 million Btu for our example house. Using the values just quoted for useful heat per gallon and our previous values for average costs of oil, gas, and electricity, the cost to replace this heat loss would be $26 for oil heat, $17 for gas heat, or $60 for electric heat.

Measure the areas of all your windows, then total up the areas for the different windows by type of glazing and material (wood or aluminum). The area will include the entire window with wood or aluminum frame. Do not include basement windows, as they will figure into the basement heat-loss section. Record your window areas in the appropriate blanks of table C-11 (Appendix C); calculate the heat loss; then calculate the gas, oil, or electricity needed to replace that heat loss.

Doors—Conduction

The R-values for primary outside doors are given in table 17. Measure the thickness of each door, and estimate the approximate percentage that is glass. If you have storm doors, add the values listed for the storm door to the R-value for the primary door to obtain the total R-value for primary and storm doors together. For example, a 1 3/4-inch-thick wood primary door with a window area of 40 percent (R = 1.7) and a wood storm door with 70 percent glass (R = 1.2) has an R-value of 2.7. If this door has an area of 21 ft^2 (a 3- × 7-foot door), annual heat loss would be 21 ft^2 × 166,000 Ddh ÷ 2.7 (R-value) = 1,300,000 Btu. The fuel needed to replace this heat loss would be 13 gallons of oil, 18 Ccf of gas, or 380 kwh of electricity. Cost of this heat loss would be $10 for oil, $6 for gas, or $23 for electricity. Again, all values are from our previously listed examples.

Now measure the areas of all your doors; record the areas in the appropriate blanks of table C-12 (Appendix C); and calculate heat loss and the gas, oil, or electricity needed to replace that loss. Do not include any basement doors—again, they will figure

TABLE 17
R-values for Outside Doors

Primary Doors: Thickness (inches)

% Glass	Wood 1	1¼	1⅜	1½	1¾	2	Insulated (Thermal door) 1¾
0	1.8	2.1	2.2	2.3	2.5	2.7	8.5
10	1.7	1.8	1.9	2.0	2.1	2.3	6.1
20	1.5	1.6	1.7	1.8	1.9	2.0	4.7
30	1.4	1.5	1.5	1.6	1.6	1.7	3.9
40	1.3	1.4	1.4	1.4	1.5	1.5	3.3
50	1.2	1.3	1.3	1.3	1.3	1.4	2.8

Storm Doors:

Metal storm door: Add 1.0 to the R-value of the primary door

Wood storm door: Add the following R-values to the R-value of the primary door

% Glass	10	20	30	40	50	60	70
R-value	1.6	1.5	1.5	1.4	1.3	1.3	1.2

into the basement heat loss section. Sliding glass doors are just large windows, so include them in the window heat-loss section.

Walls

A wall consists of a support structure, an outside covering, and an inside covering. The R-value of a wall is the sum of the R-values of these separate parts. R-values for the different support structures (including insulation) and inside and outside coverings are given in table 18. For example, a wall with 1/2-inch gypsum plasterboard as the inside covering (R = 1.1), 4 inches of brick as the outside covering (R = 1.7), and a studded support structure with 2 inches of insulation (R = 7.0) would have an R-value of 9.8 (1.1 + 1.7 + 7.). If our example house has 1,000 ft^2 of wall area (the approximate wall area for a house with 1,200 ft^2 of floor area) with this R-value, the annual heat loss would be 1,000 ft^2 × 166,000 Ddh ÷ 9.8 = 17,000,000 Btu. To replace this heat loss, an oil furnace would use about 170 gallons of oil at a cost of $128; a gas furnace would use about 240 Ccf of gas at a cost of $84; and an electric heating system would use about 5,000 kwh

TABLE 18
R-values for Outside Walls

Inside Surface (including insulating air layer):

	R-value
Gypsum plasterboard (⅜-inch)	1.0
Gypsum plasterboard (½-inch)	1.1
Gypsum plasterboard (⅝-inch)	1.2
Gypsum lath and plaster	1.1
Wood paneling (plywood)	0.8

Outside Surface (including sheathing, building paper, and insulating air layer):

	R-value
8-inch wood bevel siding, lapped	2.1
Plastic or aluminum siding	1.7
Plastic or aluminum siding over wood siding	2.4
Wood shingles	2.1
Asbestos cement shingles	1.3
4-inch brick or stone	1.7
8-inch brick or stone	2.5
Mobile home wall (metal, no sheathing)	0.2

New homes may have 1 inch of polystyrene instead of the plywood or asphalt sheathing. For this case, add 4.7 to the above R-value.

Support Structure (including insulation):

Walls with 2- × 4-inch wood studs:	R-value
No insulation	1.1
2-inch fiberglass or mineral wool	7.0
3½-inch fiberglass or mineral wool	9.6
3½-inch blown cellulose	10.6
3½-inch blown vermiculite	7.3
3½-inch blown perlite	8.7
3½-inch urea-formaldehyde foam	13.2
Walls with cinder block:	
4-inch cinder block only	1.1
8-inch cinder block only	1.7
12-inch cinder block only	1.9

If the cinder block wall has polystyrene or polyurethane insulation applied to one surface, multiply the thickness of the insulation (in inches) by the values listed in table 15 to obtain the R-value of that insulation. Add that R-value to the R-value of the wall.

Mobile home walls:

If the walls don't have two-by-four's for the support structure, find the R-value of the insulation by multiplying the thickness of the insulation (in inches) by the R-value per inch given in table 15. The resulting R-value is the R-value of the support structure.

SOURCE: *ASHRAE Handbook of Fundamentals* 1972.

at a cost of $300. (Remember that we are using a cost of 6 cents per kwh for electricity, an amount which may be a little high now, but could be actual cost in the near future.)

If you have more than one type of wall in your house, calculate the R-value for each type. Include only outside walls that face the living areas of the house, not outside walls that face unheated attics, basement walls, or crawl-space walls; these will be considered later. If you have a split-level house, include only the area above ground. Also include walls that face an unheated garage. To find the R-value of your walls, locate the appropriate R-values in table 18, then do the calculations in table C-13 (Appendix C). Then measure the areas of your outside walls for each type of wall construction, and calculate the heat loss through each of those walls in table C-14 (Appendix C).

Ceilings

The R-values for ceilings are listed in table 19. For an unheated attic above the ceiling, add the R-value for the inside surface to the R-value for the insulation to obtain the total R-value for the ceiling. If ceiling and roof are one structure, however, you must include the R-value for the roof. For example, a ceiling facing an unheated attic with gypsum lath and plaster (R = 1.1) and 6 inches of fiberglass insulation (R = 16.3) has an R-value of 17.4 (1.1 + 16.3). But if this ceiling is part of the roof structure and the roof has asphalt shingles, R-value for this ceiling is 19.1: lath and plaster (1.1) + insulation (16.3) + asphalt roof (1.7).

In table C-15 (Appendix C), calculate the R-value for each different type of ceiling, including any walls that face an unheated attic.

The roof over an attic does insulate the attic to some extent, so the average attic temperature is higher during the heating season than the average outside temperature. Actual attic temperature depends on the outside temperature and R-value of the ceiling insulation. Because heat loss through the ceiling does depend on attic temperature, we must know that temperature in order to calculate the heat loss. Table 20 gives the average difference between attic and outside temperatures as a function of attic insulation and outside temperature. Use the inside temperature closest to your average inside temperature. For example, if your

TABLE 19
R-values for Ceilings

Inside Surface (Including Surface Air Layer):

	R-value
Gypsum board, ⅜-inch	1.0
Gypsum board, ½-inch	1.1
Gypsum board, ⅝-inch	1.2
Gypsum lath and plaster	1.1
Acoustical tile over plaster	2.7
Suspended ceiling (with previous plaster ceiling)	4.3

Insulation (including ceiling joists):

	Type of Insulation			
Thickness (in inches)	*Fiberglass or Mineral Wool*	*Blown Cellulose*	*Blown Vermiculite*	*Blown Perlite*
0	0.7	0.7	0.7	0.7
1	3.0	3.5	2.3	2.8
2	5.7	6.6	4.5	5.4
3	8.4	9.6	6.6	8.0
3½	9.7	11.1	7.6	9.2
4	11.2	12.8	8.8	10.6
5	13.2	15.1	10.5	12.6
6	16.3	18.8	12.8	15.5
7	19.4	22.5	15.1	18.4
8	22.5	26.2	17.4	21.3
9	25.6	29.9	19.7	24.2
10	28.7	33.6	22.0	27.1
11	31.8	37.3	24.3	30.0
12	34.9	41.0	26.6	32.9

Roof, if Part of the Ceiling Structure (Including Sheathing and Surface Air Layer):

	R-value
Asphalt shingles	1.7
Wood shingles	2.2
Tar and stone (flat roof)	1.6
Slate	1.3
Metal	1.3

SOURCE: R-values from *ASHRAE Handbook of Fundamentals* 1972.

average inside temperature is 66° F., use 65° F. in the table. If your attic has 3 1/2 inches of fiberglass insulation (R = 10.8 from table 19) and the average outside temperature is 40° F., the temperature difference is 3° F. and the average attic temperature

Fig. 17. Examples of ceiling structures and insulation

TABLE 20
Temperature Differences Between Unheated Attic and Outside Air

Inside temperature = 65° F.

Average Outside Temp.	Ceiling R-value 1–3	4–6	7–9	10–12	13–15	16–18	20 or more
16–20° F.	20	11	8	6	5	4	3
21–25° F.	18	10	7	5	4	4	3
26–30° F.	16	9	6	5	4	3	3
31–35° F.	14	7	5	4	3	3	2
36–40° F.	11	6	4	3	3	2	2
41–45° F.	9	5	4	2	2	2	1
46–50° F.	7	4	3	2	2	1	1
51–55° F.	4	2	2	1	1	1	0
56–60° F.	2	1	1	1	1	0	0

Inside temperature = 70° F.

Average Outside Temp.	Ceiling R-value 1–3	4–6	7–9	10–12	13–15	16–18	20 or more
16–20° F.	23	12	9	7	5	4	4
21–25° F.	20	11	8	6	5	4	3
26–30° F.	18	10	7	5	4	4	3
31–35° F.	17	9	6	5	4	3	3
36–40° F.	14	7	5	4	3	3	2
41–45° F.	11	6	4	4	3	3	2
46–50° F.	9	5	4	4	3	3	1
51–55° F.	7	4	3	2	2	1	1
56–60° F.	5	3	2	1	1	1	1

Inside temperature = 75° F.

Average Outside Temp.	Ceiling R-value 1–3	4–6	7–9	10–12	13–15	16–18	20 or more
16–20° F.	25	14	9	7	6	5	4
21–25° F.	22	13	9	7	5	4	4
26–30° F.	20	11	8	6	5	4	3
31–35° F.	18	10	7	5	4	4	3
36–40° F.	16	9	6	5	4	3	3
41–45° F.	13	6	5	4	3	3	2
46–50° F.	11	6	4	3	3	2	2
51–55° F.	9	5	3	2	2	2	1
56–60° F.	7	4	3	2	2	1	1

is 43° F. (40° + 3° F. difference).[5] If your ceiling has sections with different R-values, use the R-value for the largest section.

For an example of heat loss calculation, let's consider a 1,200 ft^2 ceiling with 6 inches of fiberglass insulation (R = 17.4 from table 19) for our example house in Chicago. The average outside temperature is 40° F., so the attic temperature is 43° F. (the 3° difference from table 20 for an inside temperature of 70° F.). The difference between inside and outside temperatures is 25° F. (68° − 43° F), so the number of Degree-difference-hours for the ceiling is 25° F. × 247 heating days × 24 hours per day = 148,000 Ddh. The calculation for annual heat loss through the ceiling would be 1,200 ft^2 × 148,000 Ddh ÷ 17.4 (R-value) = 10,200,000 Btu. To replace this heat loss, an oil furnace would use about 102 gallons at a cost of $77, a gas furnace would use about 143 Ccf at a cost of $50, and an electric heating system would use about 3,000 kwh at a cost of $180.

Find the attic temperature for your situation, then use table C-16 (Appendix C) to calculate the number of Degree-difference-hours for your ceiling. This value for Degree-difference-hours will be only for a ceiling facing an unheated attic; if the ceiling and roof are one structure, the value for Degree-difference-hours would be the value calculated before using the average outside temperature (table C-10, Appendix C). Calculate the area of each type of ceiling, then calculate the heat loss through your ceiling in table C-17 (Appendix C).

Basements

Calculation of heat loss from a basement is complicated by several factors. The basement temperature is not always the same as for the rest of the house; most basement walls are underground, although parts are exposed to the outside air; and ground temperature is difficult to determine. In addition, the ground offers some insulation to the wall, but its equivalent R-value depends on many factors that are difficult to quantify. We can make some approximations, however, and thus obtain a reasonable value for basement heat loss.

First, let's consider the basement temperature. If your basement is heated to the same temperature as the rest of your house, or if you have zone heating and keep the basement at a set temperature, you already know the basement temperature. If you have a split-level house, or if part of a room in the living area is underground, consider the underground wall areas as a basement wall and the floor as a basement floor, then use the appropriate temperatures. If the basement is unheated or partially heated, the only time to measure its temperature is during the heating season. If the present time is not the heating season and you don't know the basement temperature but want to continue with the calculations, guess at the temperature and check it during the next heating season.

Measuring the ground temperature can be a little difficult, so use the values listed in table 21. Ground temperature a foot or two below the surface is approximately the average annual air temperature, and table 21 is just a list of average annual air temperatures for the degree-days given. Find the temperature that corresponds to the degree-day total for your area (recorded in table C-10, Appendix C), and use that value for your ground temperature. Then use table C-18 (Appendix C) to calculate the Degree-difference-hours for your basement. For Chicago, which has 6,300 degree-days, the ground temperature is about 50° F., so the difference between inside temperature (assumed to be 68° F.) and ground temperature is 17° F. (68° − 51° F.). The Degree-difference-hours for the basement in our example house (below ground) is 17° F. × 247 heating days × 24 hours per day = 101,000 Ddh.

The R-values for basement walls and floors are listed in table 22. For calculating heat loss through basement windows and doors, use the R-values for windows in table 16 and for doors in table 17.

Calculate the basement heat losses from the different wall and floor areas in table C-19 (Appendix C). Use the appropriate value for the Degree-difference-hours from table C-18 (Appendix C)

TABLE 21
Ground Temperatures

Degree-days	*Ground Temperature*	*Degree-days*	*Ground Temperature*
2,000	66° F.	7,000	49° F.
3,000	61° F.	8,000	45° F.
4,000	58° F.	9,000	43° F.
5,000	55° F.	10,000	40° F.
6,000	52° F.	11,000	37° F.

and R-values from table 22. Measure the wall areas with different R-values separately, and calculate the heat losses separately. The band joist, the two-by-ten or two-by-twelve that goes around the perimeter of the house, is a major heat-loss area, so measure that area carefully.

To gain some idea of the magnitudes of basement heat losses, let's assume the basement of our example house has 1,200 ft² of concrete floor, 700 ft² of 8 inch cinder-block wall below ground, 150 ft² of 8 inch cinder-block wall above ground, and 120 ft² of band joist. Heat loss through the floor would be 12,000,000 Btu (12,000 ft² × 101,000 Ddh [below ground] ÷ 10 [R-value]); heat loss through the wall below ground would be 14,000,000 Btu (700 ft² × 101,000 Ddh ÷ 5); heat loss through the wall above ground would be 12,000,000 Btu (150 ft² × 166,000 Ddh [above ground] ÷ 2); and heat loss through the band joist would be 7,000,000 Btu (120 ft² × 166,000 Ddh ÷ 2.7). Total heat loss for the basement would be 45,000,000 Btu per year. To replace this heat loss, an oil furnace would use 450 gallons at a cost of $340; a gas furnace would use 630 Ccf at a cost of $220; and an electric heating system would

TABLE 22
R-values for Basement Walls and Floors

Walls:	*Below ground*	*Above ground*	*Above ground with brick or siding*
8-inch Cinder block[a] only	5.0	2.0	2.5
8-inch Cinder block[a] with flush paneling	5.2	2.2	2.7
8-inch Cinder block[a] with furring strips and paneling	6.2	3.2	3.7
8-inch Cinder block[a] with furring strips, paneling, and ¾-inch polystyrene	7.9	4.8	5.3
8-inch Cinder block[a] with 3½-inch fiberglass insulation	16.0	13.0	13.5
8-inch Stone or brick wall	4.5	1.5	—
1½-inch wood (2-inch board, band joist)	—	—	2.7

Floors:	*R-value*
Concrete (with or without asphalt tile)	10
Concrete with carpet and foam pad	11.2
Concrete with carpet and fibrous pad	12.1

[a]For 12-inch cinder blocks, add 0.2 to the above R-values; for 4-inch cinder blocks, subtract 0.6 from the above R-values.
SOURCE: *ASHRAE Handbook of Fundamentals* 1972.

use 13,000 kwh at a cost of $780. You can see why an electrically heated home needs to be insulated!

Crawl Space

R-values for different types of floors with different thicknesses of insulation under the floor are given in table 23.

The crawl space under a house receives some heat through the floor of the house, so crawl-space temperature depends not only on ground and outside temperatures but also on the inside temperature and quality of insulation under the floor or on crawl-space walls. Use table 24 to find the temperature of the crawl space under your house. If the rooms in your house have different floor coverings, use the R-value of the floor type with the largest area. After you determine the temperature of the crawl space, calculate the number of Degree-difference-hours for the floor in table C-20, and calculate the amount of heat lost in table C-21 (Appendix C).

Let's suppose our example house has a crawl space instead of a basement, with a floor area of 1,200 ft² covered by wall-to-wall carpeting. The floor has no insulation below it, but the crawl-space walls have more than 3 inches of insulation. The crawl-space temperature is about 56° F. (from table 24 for an inside temperature of 70° F., 6,000 degree-days, and a floor R-value of 3). The difference between the inside temperature and the crawl-space temperature is 12° F. (68° − 56° F.), so the number of Degree-difference-hours for the floor is 71,000 (12° F. × 247 heating days × 24 hours per day). Annual heat loss through the

TABLE 23
R-values for Floors Over a Crawl Space

Inches of Insulation (Fiberglass or Mineral Wool)	*Hardwood Floor Over Subfloor*	*Hardwood Floor with Carpet and Foam-pad Covering*	*Hardwood Floor with Carpet and Fibrous Pad*	*Wall-to-wall Carpeting Over Subfloor*	*Linoleum Over Subfloor*
0	2.6	3.9	4.7	3.2	2.5
2	8.3	9.6	10.4	8.9	8.2
3½	12.5	13.8	14.6	13.1	12.4
6	19.3	20.6	21.4	19.9	19.2
9	27.0	28.4	29.3	27.7	26.9
12	34.1	35.5	36.4	34.8	34.0

TABLE 24
Temperature of Crawl Space

Inside temperature = 65° F.

Degree-days[a]	Floor R-value 2	3	4–7	8–12	13–17	18–22	23 or more
1,000	62(65)	62(65)	61(65)	60(65)	60(65)	60(65)	60(65)
2,000	62(63)	61(63)	60(63)	60(62)	59(62)	59(62)	59(62)
3,000	60(61)	58(60)	58(59)	57(58)	56(58)	56(58)	56(57)
4,000	58(59)	57(58)	55(56)	54(55)	53(54)	53(54)	52(53)
5,000	57(58)	55(57)	53(55)	52(53)	51(52)	51(52)	50(52)
6,000	55(56)	53(55)	51(53)	49(50)	49(50)	48(49)	48(49)
7,000	54(55)	52(54)	49(51)	47(49)	46(48)	45(47)	45(47)
8,000	52(54)	50(52)	47(49)	44(46)	43(45)	43(45)	42(44)
9,000	51(53)	48(50)	45(48)	43(44)	41(43)	41(42)	40(42)
10,000	49(52)	46(49)	43(45)	40(42)	39(41)	38(40)	38(39)

Inside temperature = 70° F.

Degree-days[a]	Floor R-value 2	3	4–7	8–12	13–17	18–22	23 or more
1,000	65(68)	64(67)	63(67)	61(66)	61(66)	61(66)	61(65)
2,000	64(66)	63(65)	62(64)	60(63)	60(63)	60(62)	59(62)
3,000	62(64)	61(62)	59(61)	57(59)	57(58)	56(58)	56(58)
4,000	60(61)	58(60)	56(58)	54(55)	54(55)	53(54)	53(54)
5,000	59(61)	57(59)	55(56)	52(54)	52(53)	51(52)	51(52)
6,000	58(59)	56(57)	53(54)	50(51)	49(50)	49(50)	48(49)
7,000	56(58)	54(56)	51(53)	48(50)	47(48)	46(48)	46(47)
8,000	54(57)	52(54)	48(51)	45(47)	44(46)	43(45)	43(44)
9,000	53(56)	50(53)	47(49)	43(45)	42(44)	41(43)	41(42)
10,000	52(54)	49(51)	45(47)	41(43)	39(41)	39(40)	38(40)

Inside temperature = 75° F.

Degree-days[a]	Floor R-value 2	3	4–7	8–12	13–17	18–22	23 or more
1,000	67(70)	66(69)	64(66)	62(67)	61(66)	61(66)	61(66)
2,000	66(68)	65(67)	63(66)	61(64)	60(63)	60(63)	60(63)
3,000	64(66)	62(64)	60(62)	58(60)	57(59)	57(59)	56(58)
4,000	63(64)	61(62)	58(59)	55(56)	54(55)	54(55)	53(54)
5,000	61(63)	58(61)	56(58)	63(55)	52(53)	51(53)	51(52)
6,000	60(62)	57(59)	54(56)	51(52)	50(51)	49(50)	48(50)
7,000	58(61)	55(58)	52(54)	49(51)	47(49)	46(48)	46(48)
8,000	57(59)	54(56)	50(52)	46(48)	44(46)	44(46)	43(45)
9,000	56(58)	52(55)	48(51)	44(46)	42(44)	42(43)	41(43)
10,000	54(57)	50(53)	46(48)	42(44)	40(42)	39(41)	38(40)

Note: Values without parentheses are for crawl spaces without wall insulation. If your crawl-space walls have 3 inches or more of insulation, use the temperature value in parentheses.
[a]Refer to table C-10 (Appendix C) for your degree-days total.

floor would be 27,000,000 Btu (1,200 ft² × 71,000 Ddh ÷ 3.2 [R-value]). To replace this heat loss, an oil furnace would use 270 gallons at a cost of $200; a gas furnace would use 380 Ccf at a cost of $135; or an electric heating system would use 8,000 kwh at a cost of $480.

Slab Floor

Heat loss through a slab floor depends on the amount of insulation around the floor perimeter, the presence or absence of heating ducts around the floor, and the floor perimeter (rather than floor area). Most homes with slab floors in the northern United States have insulation that is 12 to 24 inches wide, 1 or 2 inches thick, and installed in one of the ways shown in figure 18. Some slab floors have heating pipes or ducts around the edge that keep the floor warm; but because the warmth is around the edge, heat loss for heated floors is greater than that for unheated floors. R-values for slab floors are given in table 25. These R-values are a measure of the heat loss per foot along the floor perimeter rather than per area of the floor. The smaller R-values for heated floors reflect their larger heat loss.

Let's suppose our example house has a heated slab floor instead of a basement or crawl space. If the perimeter of the house has a length of 140 feet and has 1 inch of fiberboard insulation, annual heat loss through the edge of the floor would be 21,000,000 Btu (140 ft. × 166,000 Ddh ÷ 1.1 [R-value]). To replace this heat loss, an oil furnace would use 210 gallons at a cost of $160; a gas furnace would use 300 Ccf at a cost of $105; and an electric heating system would use 6,000 kwh at a cost of $360.

Calculate the heat loss around your floor for one heating season in table C-22 (Appendix C). If you are not sure if your floor is insulated, consult the contractor who built the house.

TABLE 25
Effective R-values for Slab Floors

Insulation Thickness:	2 inches		1 inch		None
	Polystyrene	*Fiberboard*	*Polystyrene*	*Fiberboard*	
Heated	4.5	2.2	2.2	1.1	0.8
Unheated	5.5	3.0	3.0	1.5	1.2

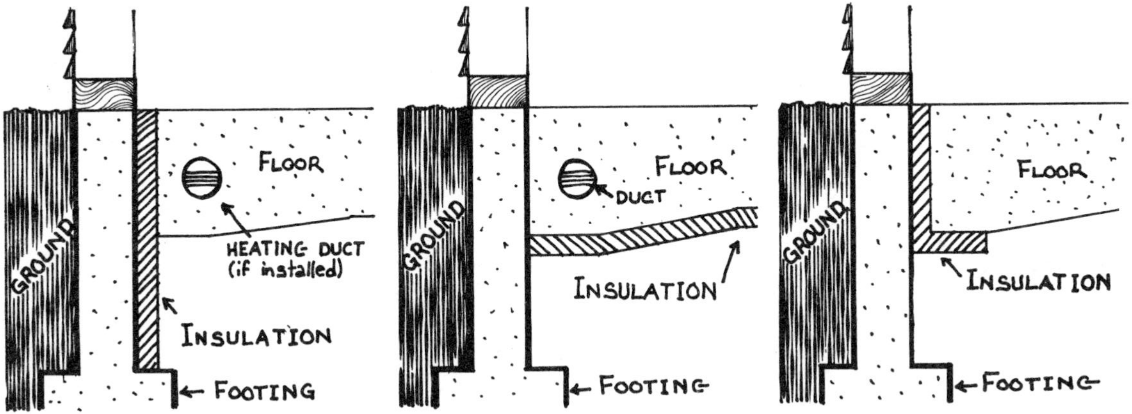

Fig. 18. Arrangements of insulation around a slab floor

Infiltrative Heat Losses

Previous calculations for conduction heat losses are based on values that are easy to measure with reasonable accuracy (areas, R-values, and temperatures), so these calculated values should closely represent the amounts of actual heat loss. To calculate the amount of air infiltration through cracks with reasonable accuracy, however, you need information that is difficult, if not impossible, to obtain: the width and length of the cracks around your windows and doors and the average wind speed outside your house. Not only are most door and window cracks (plus cracks in walls and ceilings) impossible to measure when found, but we probably can't even see most cracks that do exist. And wind speed, although recorded at local weather stations, can vary greatly between locations, depending on the local geography and where your house is situated. Therefore the best you can do is to use some method of calculation that can give at least some rough approximation of the infiltration rate.

The air-change method can give us this information. Infiltration of air into a room generally depends on the number of outside walls with windows or doors. For example, a room with one outside wall and a window has approximately 1 air change per hour if the window has no storm window or weather stripping. If the room is 10 feet wide, 10 feet long, and 8 feet high (volume = 800 ft^3), the infiltration rate would be 800 cubic feet per hour. Table 26 gives the approximate number of air changes per hour for rooms with various numbers of walls with windows or outside doors. These values are reasonably accurate for most houses; but

TABLE 26
Air-Change Rates for Rooms

Number of walls with windows or exterior doors	*Windows, doors No storm or weather stripping*	*Windows and doors With storm or weather stripping*
0	½	⅓
1	1	⅔
2	1½	1
3	2	1⅓

houses with exceptionally good or exceptionally poor seals around windows and doors may have air-change rates considerably different from the table values.

The amount of annual heat loss for a house is the product of the infiltration rate (cubic feet per hour), the number of hours in the heating season, and the energy needed to heat each cubic foot of infiltrating air. The amount of energy needed to heat the infiltrating air is found by multiplying the temperature rise of the air by 0.018 Btu of energy needed to heat one cubic foot of air by 1° F. Then, if we combine the number of hours in a heating season with the temperature rise of the infiltrating air, we end up with the number of Degree-difference-hours for the house. Thus the annual heat loss for a house is the product of the cubic-feet-per-hour infiltration rate, the number of Degree-difference-hours for the house, and 0.018 Btu per cubic foot of air.

For our example house, let's assume that every room has two walls with a window or outside door; and that the windows and doors are weather stripped; so the air-change factor (from table 26) is 1 air change per hour. The house has a floor area of 1,200 ft^2 and a ceiling height of 8 feet, so the volume is 9,600 ft^3 (1,200 ft^2 × 8 ft). The cubic-feet-per-hour infiltration rate is the air change factor multiplied by the volume of the house; so for this house, the infiltration rate would be 9,600 ft^3 per hour (9,600 ft^3 volume × 1 air change per hour). The annual heat loss would be 29,000,000 Btu (9,600 ft^3 per hour × 166,000 Ddh × 0.018 Btu per cubic foot of air). To replace this heat loss, an oil furnace would use 290 gallons at a cost of $220; a gas furnace would use 410 Ccf at a cost of $145; and an electric heating system would use 8,500 kwh at a cost of $510.

Calculate the infiltration heat loss for your house in table C-23 (Appendix C). Group the rooms according to the number of walls with windows or doors, and find the total area of those rooms. Then use the appropriate factors from table 26 to find the infiltration rate in cubic feet per hour and the energy needed to heat the infiltrating air.

Hot-air Ducts in Unheated Areas

If you have forced hot-air heat and some of the air ducts pass through an unheated area of your house (such as attic or crawl space), these ducts lose some heat to this area. To calculate this heat loss, you must know the surface area of the exposed ducts, the R-value of the ducts, the temperature of the hot air in the ducts, and the total amount of time that your furnace is operating during the heating season.

To find the amount of time that your furnace is operating during the heating season, all you need to know is the total amount of energy used during the heating season and the input of the furnace in gallons of oil per hour, cubic feet of gas per hour, or electric kilowatts. The total energy used during a heating season is recorded in table C-8, and the input of your furnace is recorded in table C-7 of Appendix C. Record these values in table C-24 of Appendix C, then calculate the total number of hours that your furnace is operating during the heating season.

To find the area of the ducts and the temperature inside the ducts requires a little more effort, so fetch your thermometer and ruler and start exploring your attic or crawl space. Measure the distance around the duct (all four sides if the duct is rectangular) and its total length. If the duct narrows along its total length, measure the distance around each section and the length of each section, then calculate the total area of each section. The area is the length of the duct section multiplied by the distance around the duct. Add up the areas of each section, and record the total area in table C-25 (Appendix C).

To measure the temperature, place your thermometer directly on the hot-air register located about one-third of the distance from the furnace to the farthest register from the furnace. The air temperature at this register should represent the average temperature of the air in the duct. Be sure the thermometer is in place long enough to record the temperature of the air in the register. If the furnace turns off before the thermometer reading reaches the hot-air temperature, you may need to turn up your thermostat a little to be sure the thermometer records the proper reading. (Don't forget to reset the thermostat when you're finished.) Record this temperature in table C-25 (Appendix C).

Amount of heat loss through the duct walls depends on the insulation, if any, around the duct. The R-value of the duct itself is 0.8, almost entirely from the insulating layer of outside surface air—the duct metal offers little resistance to heat conduction. R-value of a duct with insulation equals the R-value of the insulation.

Amount of annual heat loss through the duct walls is the duct area multiplied by the temperature difference between inside and outside of the duct multiplied by the number of hours the furnace is operating, then divided by the R-value of the duct. If the ducts in our example house are uninsulated, have an area of 100 ft^2, and have a hot-air temperature of 130° F.; if the attic temperature is 43° F. and the furnace is operating for 700 hours during the heating season; then the annual heat loss through the duct walls would be 7,600,000 Btu (100 ft^2 × 87° F. × 700 hrs. ÷ 0.8). To replace this heat loss, an oil furnace would use 75 gallons at a cost of $55; a gas furnace would use 110 Ccf at a cost of $40; and an electric furnace would use 2,200 kwh at a cost of $130.

Use table C-25 (Appendix C) to find the R-value of your ducts, then calculate the annual heat loss through your duct walls.

Adding It All Up—With Ifs, Ands, and Buts

In table C-26 (Appendix C), record the amount of energy needed to replace the heat losses from the different parts of your house. Then add them together to find the total amount of energy that must be supplied to replace these heat losses.

Before you compare the total of your heat losses with the amount of energy you actually use to heat your house, let's consider the accuracy of your heat-loss values. All calculations for conductive heat losses are based on ideal conditions. If your insulation is not properly or uniformly installed, the R-values for your walls and ceilings are not exactly correct. The R-values used to calculate the heat losses through windows and doors are based on a 10-mph wind, so if some of your windows are exposed to a higher wind, their heat losses are more than you calculated. Conversely, windows sheltered from the wind have lower heat losses than you calculated. Generally you can probably assume that your calculated conduction heat losses are within 10 to 20 percent of your actual heat losses.

As for your infiltration heat losses, remember that those calculations depend on the number of walls with windows or doors and whether your house has weather stripping or storm windows and doors. Actual infiltration, however, depends on how many windows and doors you have, the quality of their fit, the wind speed for your area, the quality of your weather stripping and storm doors or windows, and many other factors that are hard to quantify. Therefore the best you can expect from this calculation is a 20 to 50 percent accuracy.

Now we are ready to compare numbers. If your calculated heat losses (table C-26, Appendix C) are less than the amount of energy you actually use for heat (table C-8, Appendix C), you have problems. First, check your figures to be sure your calculations are correct. If correct, then check your house carefully to locate any possible areas of additional heat losses. If you can't find any more sources of possible heat loss, your infiltration heat losses are probably much higher than you calculated. Check for any large openings around your house that are not readily visible, and check for exceptionally large cracks around the fit of your windows and doors.

If your calculated heat losses are greater than the amount of energy you actually use for heat, we're in business. Generally your calculated heat losses (table C-26, Appendix C) should be from 20 to 50 percent more than the amount you use for heat (table C-8, Appendix C). But shouldn't the amount of heat energy produced by your heating system be equal to the heat losses? It would be if your heating system was the only source of heat for your house—but it isn't; people, sunlight, and your electrical appliances contribute significant amounts of heat to your house—but their exact contributions, especially from solar energy, are difficult to calculate.

Every kilowatt hour of electrical energy used in the home ends up as heat. A light bulb is just as efficient as an electrical baseboard heater—each kilowatt hour consumed by the light bulb produces 3,413 Btu of heat. Most of the energy consumed by a light bulb becomes heat directly (the bulb is hot); but even energy in the form of light is absorbed by the walls, floors, and objects of the room and is also converted to heat. Similarly the electrical energy used to operate a television or radio ends up as heat (these appliances become warm with use).

But heat from electrical appliances contributes to the house heat only if the appliances are in the house and the heat produced by them remains in the house. Outdoor lights contribute nothing to the heat in the house, and an electric clothes dryer vented to the outside contributes little. Because most of the hot water used in the home goes down the drain without leaving much heat behind, most of the electrical energy used to heat the water is also lost from the house. Likewise most of the electrical energy used by a dishwasher is used to heat the water, and this heated water drains out with only a little of it remaining to heat the house.

To gain some idea of the amount of heat contribution by electrical appliances, let's consider a family that uses 300 kwh per month for appliances other than a clothes dryer, hot-water heater, dishwasher, and outdoor lighting. The 300 kwh contributes 1,020,000 Btu to the house, which is the same amount of heat produced by 10 gallons of oil or 15 Ccf of gas in a furnace. For a seven-month heating season, this heat contribution amounts to about 70 gallons of oil or 105 Ccf of gas. For homes with electrical resistance heat, these 300 kwh produce the same heat as 300 kwh used by the heating system.

The only contribution to home heat from gas appliances is the heat used for cooking. Heat from clothes drying, hot-water heating, and outside gas lamps is either outside, vented outside, or lost down the water drain.

The heat energy produced by most adult humans is between 400 and 600 Btu per hour, depending on the person's activities; so an adult who is home 16 hours per day can contribute from 6,000 to 10,000 Btu per day to the heat in the house. For a period of one month, the heat produced by one person is approximately equivalent to the heat produced by 2 gallons of oil, 3 Ccf of gas, or 70 kwh of electricity. The heat produced by a family of four during a seven-month heating season would equal the amount of heat produced by about 60 gallons of oil, 90 Ccf of gas, or 1,800 kwh of electricity. (So the next time someone says you are worthless, tell him how much you are actually worth!)

The contribution of solar energy can range from a few percent to 20 or 30 percent of the total home heating requirements. The exact amount depends on size and arrangements of the windows, type and color of the outside walls, quality of insulation, amount of shading of the house, latitude, and number of cloudless days per heating season. Because these factors are different for each house, the amount of heat gain from solar energy is difficult to calculate for any particular house.

Humidifiers and Fireplaces—Do They Really Help?

We often read that a humidifier helps to reduce the energy needed to heat a house or that a fireplace is a good extra source of heat. Do they really help? The answer is yes, no, and maybe, depending on several factors.

Let's look at fireplaces first (and about all that many fireplaces are good for is just "looking at"). The only heat supplied to a room by a normal fireplace comes from the heat radiated, which is about 10 percent of the heat generated by the fire. Although 10 percent is better than nothing, you must also consider that warm air from the room goes up the chimney with the smoke. Such heat loss plus the warm air still going up the chimney after the fire is out but still smoking can just about cancel the amount of room heat gained from the fire. So if you have a fireplace with no effective way to transfer its heat to the room air, that fireplace provides little if any heat to your house. In many cases, more heat is actually lost than gained.

To be an effective source of heat, a fireplace needs some sort of heat exchanger to heat the room air. The best type of heat exchanger is built into the fireplace, as shown in figure 19. The room air enters at the bottom, picks up heat from the fireplace walls, then exits at the top. A fireplace with a heat exchanger of this type can produce a significant amount of heat, but this system must be installed when the fireplace is built.

The most common type of heat exchanger for finished fireplaces is a tubular grate, also shown in figure 19. Room air enters the bottom, becomes heated in the tube, and exits at the top. While this arrangement is not as effective as the installed heat exchanger, it can help heat the house to some extent.

A humidifier reportedly saves energy because people supposedly feel warm at a lower room temperature if the relative humidity is increased. This is not true—the only factor that determines a person's warmth is the degree of air temperature. Increased humidity does have some positive aspects, however. It reduces static electricity, shrinkage of wood and textiles, chapping of skin and lips, and discomfort of dry throats.

A humidifier uses energy, however, to increase air humidity. Most humidifiers don't use heat directly to evaporate the water. Instead, a small motor moves a wide belt through a pan of water; as air circulates past the wet belt, the water evaporates. But the evaporating water removes about 1,000 Btu of heat from the air for each pound (about a pint) evaporated, and the heating system must then replace this heat. To maintain a 30-percent relative

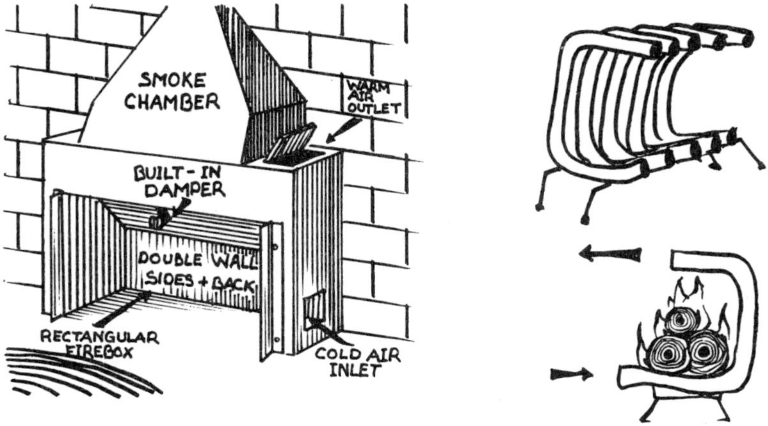

Fig. 19. Fireplace with effective heat exchangers: *left,* built-in; *right,* tubular grate

humidity inside when the outside temperature is 30° F., a humidifier must evaporate about 2 lb. of water per hour. This water vapor leaks out of the house as an additional convective heat loss. To evaporate these 2 lb. per hour (48 lb. per day—about 6 gallons of water), your heating system must produce an additional 48,000 Btu per day which requires 0.7 Ccf of gas, 0.5 gallons of oil, or 14 kwh of electricity. Over a seven-month heating season, a humidifier could use up to 150 Ccf of gas, 100 gallons of oil, or 3,000 kwh of electricity.

Fuel Saving Devices for Your Furnace

Two fuel saving devices currently on the market may or may not work for you. One is a heat exchanger on the flue pipe of the furnace; the other is a flue damper. Both of these devices can save energy, but you should know some important facts before you rush out to buy one.

The flue heat-recovery device is a heat exchanger that can extract heat from the flue gases, but flue temperature must be above 300° F. for proper operation. If your flue temperature is less than 300° F., the heat recovery unit will cool the flue gas to the point where condensation can occur in the chimney—and with condensation comes corrosion. So consult the people who service your furnace or your gas utility to see if you can install a flue heat-recovery system.

With the flue pipe constantly open, warm air from inside the house flows up the chimney when the furnace is not in operation. You can purchase a device that closes the flue pipe when the furnace is off to save this heat loss (the device opens the pipe when the furnace is operating to allow flue gases to escape up the chimney). This device is fine if it works perfectly every time—but if it malfunctions only once (that is, doesn't open the pipe when the furnace turns on), results could be disastrous: either the flue gases would enter the house, the area around the furnace would get hot enough to start a fire, or both. If you want to reduce the air flow out the flue pipe, you should reduce air infiltration into the house (caulk and weather-strip), then provide an independent air source for your furnace from outside. The furnace would then draw the air from outside rather than through the entire house.

Conservation Measures

The following paragraphs suggest several energy conservation measures, for many of which you can calculate the amount of expected money and energy savings. Because some of the calculations are long (generally requiring you to repeat earlier calculations), several of the sections contain a "rough estimate" calculation. These "rough estimate" values are based on average figures, enabling you to gain some idea of your expected savings even if you haven't done all the previous calculations for heat losses. Conditions for these rough estimates may or may not apply to your own house; however, if they do apply, the amount of savings calculated from these values should be fairly close to your actual savings. If the conditions listed don't apply to your house, then the calculated savings will not necessarily be the amount that you would actually save, although they should give you some idea of your savings. To arrive at a more accurate value for your expected savings, you must do the calculations listed under the "Energy Savings: Better Value" headings that follow.

Heat loss by conduction, as we have seen, depends on the temperature of inside and outside air; on wall, ceiling, and other areas and their R-values; and on length of the heating season.

Thus the only way to reduce conductive heat loss is to change the areas, the R-values, the temperature of inside and outside air, or the length of the heating season. Let's assume that we can't shrink the house to reduce the area and that we can't increase the outside temperature or decrease the length of the heating season without moving the house to a warmer climate. Thus we are left with only two possible ways to reduce conduction heat losses—reduce the inside air temperature and increase the R-values of the insulation around the house. To reduce infiltration heat losses, we can only reduce the volume of infiltrating air and reduce the inside temperature.

Reduce Inside Temperature

Lower the inside temperature to 68° F. during the day and to 64° F. at night. If you feel comfortable at lower temperatures, reduce them further. This is the easiest and, needless to say, least expensive way to save energy; but if you're accustomed to a warm inside temperature, these cooler temperatures may be uncomfortable. Most people can adjust to a lower temperature, however. Most studies analyzing the effect of cooler temperatures on susceptibility to catching colds and other illnesses conclude that the lower temperatures do not affect health; some studies, in fact, indicate that lower temperatures are healthier. Small children can take lower temperatures better than adults. A five-year-old child produces about twice the heat per body area of an adult and therefore has no trouble keeping warm in the cooler temperatures—kids don't even get goose bumps when they run around in their birthday suits.

Energy Savings: "Rough Estimate" Decide what inside temperatures (day and night) you want, calculate your average inside temperature (as in table C-9, Appendix C), then find the difference between new and old inside temperatures. This difference is your temperature reduction. Refer to table 27 for the expected percent savings for the degree-days of your area (table C-10); then you can calculate your expected fuel and dollar savings in table C-27 (Appendix C).

Energy Savings: "Better Value" To obtain a more accurate value for your savings, you must go back to the sections on heat-loss calculations and recalculate heat losses for your new temperature setting, starting with tables C-9 and C-10 (Appendix

TABLE 27
Percent Savings from Inside Temperature Reduction

Degree-days	Temperature Reduction (°F.) 1	2	3	4	5	6	7	8
	%	%	%	%	%	%	%	%
1,000	8	14	20	26	32	38	43	48
2,000	7	12	16	21	25	30	34	38
3,000	5	9	12	16	19	23	26	29
4,000	4	8	11	15	18	21	24	27
5,000	3	7	10	14	17	20	22	24
6,000	3	6	9	12	14	17	19	22
7,000	2	5	7	10	13	16	18	21
8,000	2	3	7	9	11	14	16	19
9,000	2	4	6	9	10	13	15	17
10,000	2	4	6	8	10	12	13	15

C) for the new average inside temperature, the new numbers of heating days, and the average outside temperature for those heating days. Then calculate the new value for the Degree-difference-hours, and continue through the calculations for heat losses from various parts of the house. The difference between the heat loss before decreasing your temperature and after decreasing it is the energy saved. As you work through the various sections, record your new heat-loss data in table C-28 (Appendix C). As no new tables are provided for these new calculations, follow the outline of tables in the heat-loss sections but do your calculations on another piece of paper.

Insulate Walls, Ceilings, or Floors

Install insulation or increase the thickness of insulation in walls, ceilings, and floors (above a crawl space). The minimum recommended thickness of insulation for a ceiling is 6 inches of fiberglass (R-19) or its equivalent. Houses in regions with 6,000 degree-days or more and homes with electric heat should have at least 9 inches. If you plan to add insulation to your attic, find the cost of installing different thicknesses, then calculate the energy and money savings for each thickness and decide which is the most economical for you. Depending on cost of insulation, cost of your heating fuel, and severity of the winter in your area, 6 inches of insulation installed in an attic with no insulation can save enough energy in one to four years to pay for the insulation. Even adding 6 inches on top of 3 1/2 inches of existing insulation can save enough energy in five to ten years to pay for itself.

Most homeowners can install attic insulation themselves. The job may be hot and dirty, but it's not too difficult. Fiberglass insulation comes in rolls and batts (4-foot-long pieces) and fits between the ceiling joists. When purchasing the insulation, be sure to know the width you need, either 15- or 23-inch width for 16- and 24-inch joist centers, respectively. Take the necessary time to lay the insulation properly so it fits tightly between the joists to reduce air movement around the insulation. Don't block any outside vents that allow proper ventilation of the attic, and don't block the exhausts from bathroom or kitchen fans. Be sure to wear gloves and a face mask to protect your skin and lungs from the fiberglass.

If your attic space is too tight for free movement, you may want to consider blown-in insulation. You can rent the blower apparatus from your local insulation dealer or have a contractor install the insulation for you. If you have any questions about proper insulation, check with your insulation dealer, a contractor, utility, or a home do-it-yourself book.

If you insulate your crawl space, be careful of your water pipes. Heat coming through the floor is the only heat there is for the crawl space; so if you insulate the floor, the crawl-space temperature could drop below freezing during a cold period of the winter and freeze the pipes—so be sure that all water (and drain) pipes are between the insulation and the floor. If the pipes are fastened to the bottom of the floor joists, you can fabricate a covering for them, but every pipe must have direct access to heat from the floor. Don't wrap a pipe completely with insulation—when the water isn't flowing, the pipe has no heat source. Even with lots of insulation, it could cool enough to freeze. If you can't insulate the floor because of water pipes, the next best action is to insulate the walls of your crawl space. For this job you should consult a book or person who can show you the proper method of installation.

Walls are another problem entirely. Because walls are often closed on both sides, you can't easily slip insulation between inside and outside walls. Walls with 3 1/2 inches of insulation are fully insulated, and no more can fit between them. If your walls have no insulation, you will need to have a qualified contractor blow insulation between the walls. Small openings are made along the outside wall, and the insulation is blown through them into the gap between the walls. A good contractor can patch the series of holes so they are hardly noticeable. If your walls have only 2 inches of insulation, you may want to consult a contractor to see if any more can be blown into the remaining air gap. With uninsulated walls and water pipes going up the outside wall, be sure that no insulation lies between the inside wall surface and the pipe. Only the heat from inside the house keeps the pipe from freezing in the winter.

Some older homes may have solid brick or stone walls with plaster applied directly on the inside surface. These walls obviously have no space for insulation between "inside" and "outside" coverings. To add insulation, you must install either two-by-four or two-by-two studs, or perhaps just 1-inch furring strips to the inside wall; install insulation (fiberglass or mineral wool for two-by-four studs, polystyrene or polyurethane for two-by-two studs or furring strips), then cover the insulation with paneling or plaster.

Not only does added insulation reduce conduction heat loss, but you will feel more comfortable even at a lower inside temperature. A poorly insulated or noninsulated wall or ceiling is considerably colder than the room air. When outside temperature is 30° F. and inside temperature is 70° F. in an uninsulated home, wall and ceiling temperature is only about 63° F.—so walls and ceiling absorb more radiation energy than they radiate themselves. While your body radiates heat to walls and ceiling, walls and ceiling don't return the same amount of radiation, so you feel cold even if the air temperature is 70° F. or more. A well-insulated wall or ceiling measures about the same temperature as the room air, thus returning almost an equal amount of radiation as you send to it.

Energy Savings: "Rough Estimate" Table 28 gives the Btu saved per square foot during a heating season for the insulation changes noted. The values for ceilings and crawl spaces are for fiberglass or mineral wool insulation, and values for the walls are for the insulations given. If you have a different type of ceiling insulation, determine the present and final R-values for that insulation (multiply the insulation thickness by the R-value per inch from table 15). Locate your approximate expected savings in the column for the R-value closest to your initial and final R-values.

The values in table 28 are based on an inside temperature of 68° F., so if your inside temperature is less than that, your savings will be less; and if your inside temperature is more than 68° F., your savings will be greater than those shown. Find heat-loss reduction for the planned increase of insulation for your area degree-days in table 28; then calculate your fuel and money savings in table C-29 (Appendix C).

For example, if you live in an area with 6,000 degree-days and you increase your ceiling insulation from 3 to 6 inches, you would save about 4,200 Btu per square foot of ceiling. For a 1,000-square-foot ceiling, reduced heat loss per year would be 4,200,000 Btu, which would mean savings of about 40 gallons of fuel oil, 60 Ccf of gas, or 1,250 kwh of electricity. Dollar savings per year would be about $25 for gas or oil and about $65 for electricity.

Energy Savings: "Better Value" Now return to our previous sections on heat losses through walls, ceilings, and crawl spaces, and recalculate heat losses with the proposed added insulation. The amount of savings is the difference between the "before" and "after" values. If you plan to reduce your inside temperature as suggested in the previous section, you must calculate the "before" and "after" heat losses at your new inside temperature. The "Before" value for this calculation is the "After" value in table C-28 (Appendix C).

Calculate your expected savings in table C-30 (Appendix C).

Insulate Your Basement

Major heat loss in a basement is through the portion of the wall above the outside ground level. Insulate the band joist with 6 inches of insulation; and if your basement is not decorated, install insulation down to a foot or two below ground level. If you plan to panel the walls, install 3/4-inch of plastic foam insulation behind the paneling. Because plastic foam insulations can produce toxic fumes and melt rather than burn during a fire, always cover them on the inside of the house.

Energy Savings: "Rough Estimate" Table 29 gives the Btu saved per heating season for the insulation added as indicated. These values are for a basement with 8-inch cinder-block walls, a 2-inch-thick band joist with brick, stone, or siding on the outside surface, and a basement temperature of 68° F. For your area degree-days and the planned added insulation, find the appro-

TABLE 28
BTU per Square Foot Saved by Adding Insulation

Ceilings:

Degree-days	Inches of Fiberglass Insulation Thickness Increased From: 0 to 3 (R-0 to R-11)	0 to 6 (R-0 to R-19)	0 to 9 (R-0 to R-30)	3 to 6 (R-11 to R-19)	3 to 9 (R-11 to R-30)	6 to 9 (R-19 to R-30)	6 to 12 (R-19 to R-36)
1,000	8,000	8,900	9,400	700	1,300	500	800
1,500	12,000	12,000	13,000	1,200	2,000	700	1,100
2,000	15,000	16,000	17,000	1,700	2,600	900	1,400
2,500	17,000	19,000	20,000	2,000	3,200	1,200	1,800
3,000	20,000	23,000	24,000	2,300	3,800	1,500	2,300
3,500	22,000	26,000	28,000	2,600	4,300	1,700	2,700
4,000	25,000	30,000	32,000	3,000	4,900	1,900	3,100
4,500	28,000	32,000	35,000	3,300	5,300	2,100	3,400
5,000	33,000	35,000	38,000	3,600	5,900	2,300	3,700
5,500	35,000	37,000	40,000	3,900	6,400	2,500	4,000
6,000	37,000	40,000	43,000	4,200	7,000	2,800	4,400
6,500	40,000	44,000	48,000	4,800	7,900	3,100	4,700
7,000	44,000	49,000	53,000	5,400	8,800	3,400	5,000
7,500	47,000	54,000	57,000	5,800	9,500	3,600	5,400
8,000	51,000	58,000	61,000	6,300	10,200	3,900	5,800
8,500	52,000	59,000	62,000	6,500	10,500	4,000	6,000
9,000	53,000	60,000	64,000	6,700	10,900	4,200	6,200
9,500	54,000	61,000	65,000	6,800	11,100	4,400	6,500
10,000	55,000	62,000	67,000	6,800	11,400	4,600	6,800

Crawl Space:

Degree-days	Inches of Fiberglass Insulation 0 to 3	0 to 6	0 to 9	Thickness Increased From: 3 or more to walls;[a] no floor insulation
1,000	1,400	1,900	2,200	3,000
1,500	2,100	2,700	3,000	3,000
2,000	2,700	3,500	3,900	3,000
2,500	3,900	4,800	5,500	3,000
3,000	5,200	6,200	7,000	3,000
3,500	6,700	8,000	8,900	3,200
4,000	8,100	9,700	10,800	3,400
4,500	9,100	10,900	12,100	3,600
5,000	10,000	12,000	13,400	3,800
5,500	10,800	13,200	14,200	3,900
6,000	11,500	14,400	15,900	4,100
6,500	13,500	16,600	19,500	4,200
7,000	15,600	18,800	21,000	4,400
7,500	17,100	20,500	22,900	4,400
8,000	18,500	22,200	24,700	4,400
8,500	19,100	23,000	25,500	4,400
9,000	19,700	23,700	26,400	4,400
9,500	20,000	24,500	27,100	4,400
10,000	20,300	25,200	27,800	4,400

TABLE 28 (*Continued*)

Walls:

Degree-days	Thickness Increased from 0 to 3½ Inches of Following Insulation: Blown Vermiculite	Blown Perlite	Blown Cellulose	Urea Foam
1,000	3,600	3,900	4,200	4,600
1,500	5,400	5,900	6,300	7,000
2,000	7,100	7.800	8,500	9,200
2,500	9,000	10,000	11,000	12,000
3,000	11,000	12,000	13,000	14,000
3;500	12,000	14,000	15,000	16,000
4,000	14,000	16,000	17,000	18,000
4,500	15,000	17,000	19,000	20,000
5,000	17,000	19,000	21,000	22,000
5,500	19,000	21,000	23,000	24,000
6,000	21,000	23,000	25,000	27,000
6,500	22,000	25,000	27,000	29,000
7,000	24,000	27,000	29,000	31,000
7,500	26,000	29,000	31,000	33,000
8,000	28,000	31,000	33,000	36,000
8,500	29,000	32,000	34,000	37,000
9,000	31,000	33,000	36,000	39,000
9,500	32,000	34,000	37,000	41,000
10,000	33,000	36,000	39,000	43,000

[a]These values are Btu saved per square foot of floor area, not per square foot of insulation installed on the walls. Values are based on insulation for all four walls.

priate value in table 29, then calculate your expected savings in table C-31 (Appendix C).

For example, if you live in an area with 6,000 degree-days and add 6 inches of fiberglass insulation to the band joist, you would save 50,000 Btu for every foot of 12-inch band joist insulated. If your house has 100 feet of band joist, reduced heat loss per year would be 5,000,000 Btu. This heat-loss reduction would mean savings of about 50 gallons of oil, 70 Ccf of gas, and 1,500 kwh of electricity—dollar savings of about $30 for oil or gas and $75 for electricity.

Energy Savings: "Better Value" Calculate your expected heat loss with insulation installed and subtract that figure from your previously calculated heat loss (table C-19, Appendix C). If you are planning to reduce your basement temperature, you must calculate the "Before" and "After" heat losses at the new base-

TABLE 29
BTU Saved by Adding Basement Insulation

Add the following insulation:

Degree-days	6-inch Fiberglass to Band Joist Size of Joist: 10 inches	12 inches	3½-inch Fiberglass to Wall: Above Ground	Below Ground	¾-inch Polystyrene to Wall: Above Ground	Below Ground
1,000	7,000	8,000	10,000	—	7,000	—
1,500	10,000	12,000	15,000	—	10,000	—
2,000	14,000	17,000	20,000	1,000	14,000	1,000
2,500	17,000	20,000	26,000	3,000	17,000	2,000
3,000	21,000	25,000	31,000	4,000	21,000	3,000
3,500	23,000	30,000	36,000	6,000	25,000	4,000
4,000	26,000	35,000	42,000	8,000	29,000	5,000
4,500	29,000	37,000	45,000	9,000	31,000	5,000
5,000	33,000	40,000	49,000	10,000	33,000	6,000
5,500	37,000	45,000	54,000	12,000	37,000	7,000
6,000	41,000	50,000	60,000	14,000	41,000	8,000
6,500	44,000	53,000	64,000	15,000	44,000	9,000
7,000	47,000	57,000	69,000	16,000	47,000	10,000
7,500	50,000	61,000	74,000	17,000	50,000	10,000
8,000	54,000	65,000	79,000	18,000	53,000	11,000
8,500	56,000	68,000	83,000	19,000	55,000	11,000
9,000	59,000	71,000	87,000	20,000	58,000	12,000
9,500	61,000	74,000	90,000	21,000	60,000	13,000
10,000	64,000	78,000	94,000	23,000	63,000	14,000

NOTE: Values listed for the band joist are Btu per linear foot of band joist; values for the wall are Btu per square foot.

ment temperature. The "Before" value in this case will be the "After" value calculated in table C-28 (Appendix C).

Calculate your expected savings in table C-32 (Appendix C).

Insulate Your Slab Floor

If your home has a slab floor with no insulation around the floor perimeter, install plastic foam insulation around the outside surface of the foundation. You cannot see any insulation installed between foundation and floor or under the floor, so you will have to ask the contractor who built the house if any floor insulation was installed. To install the insulation, dig around the foundation to a depth of 24 inches or to the bottom of the foundation—whichever comes first—and apply the insulation to the foundation surface. Check with your insulation dealer, contractor, utility, or a do-it-yourself manual for the proper method of installation.

Energy Savings: "Rough Estimate" Table 30 gives the Btu saved per heating season for insulation added to the perimeter of a slab floor. These values are for polystyrene insulation added to a depth of 24 inches. If your foundation doesn't go down that deep, your savings will be less than the values indicated. For your area degree-days, find the appropriate value in table 30, then calculate your expected savings in table C-33 (Appendix C).

Energy Savings: "Better Value" Calculate the expected heat loss with the insulation installed, then subtract that heat loss from your previously calculated heat loss (table C-22, Appendix C). If you plan to reduce the inside temperature, you must calculate the "Before" and "After" heat losses at the new inside temperature. The "Before" value in this case will be the "After" value calculated in table C-28 (Appendix C).

Calculate your expected savings in table C-34 (Appendix C).

TABLE 30
BTU Saved per Linear Foot by Adding Insulation to a Slab Floor Perimeter

Add the following thickness of polystyrene insulation:

Degree-days	Heated Floor 1-inch	Heated Floor 2-inch	Unheated Floor 1-inch	Unheated Floor 2-inch
1,000	21,000	28,000	13,000	17,000
1,500	32,000	41,000	20,000	26,000
2,000	43,000	55,000	27,000	35,000
2,500	54,000	69,000	34,000	44,000
3,000	65,000	84,000	41,000	53,000
3,500	75,000	97,000	47,000	61,000
4,000	85,000	110,000	53,000	69,000
4,500	94,000	121,000	59,000	76,000
5,000	103,000	133,000	65,000	84,000
5,500	115,000	148,000	72,000	94,000
6,000	127,000	164,000	80,000	102,000
6,500	136,000	176,000	86,000	112,000
7,000	146,000	188,000	92,000	120,000
7,500	156,000	202,000	98,000	128,000
8,000	167,000	215,000	105,000	137,000
8,500	175,000	225,000	110,000	143,000
9,000	183,000	236,000	115,000	149,000
9,500	190,000	246,000	119,000	155,000
10,000	198,000	256,000	124,000	162,000

Insulate Your Heating Ducts

Cover your exposed heating ducts with at least 3 1/2 inches of fiberglass or mineral wool insulation. Don't pull the insulation too tight, or its thickness and R-value will decrease. Also inspect your ducts to be sure no holes exist that let hot air escape.

Energy Savings Because heat loss from a heating duct in an uninsulated area depends on several variables that are hard to typify for an "average" home, you must complete the entire calculation to find your own expected savings. Calculate your expected heat loss with insulation installed, then subtract that heat loss from your previously calculated heat loss (table C-25, Appendix C).

Calculate your expected savings in table C-35 (Appendix C).

Install Storm Windows and Doors

Depending on the region of the country, storm windows and doors will pay for themselves with energy saved in three to ten years. Even if you already have double-glazed (thermopane) windows, savings through storm insulation can be significant. Storm windows also reduce the infiltration of outside air.

And, as with insulated walls and ceilings, the addition of storm windows will make you feel warmer. For an outside temperature of 30° and an inside temperature of 70° F., the inner surface temperature of the glass in a single-glazed window is only 40° F.; but inner surface temperature with a storm window is 55° F. If the window has triple glazing, inside glass temperature is 60° F. The glass not only radiates more heat to the room at this higher temperature, but less water vapor condenses on the windows, and room air can maintain a higher relative humidity.

Energy Savings: "Rough Estimate" Table 31 gives the Btu saved per heating season by adding storm windows and doors. The door values are based on a 1 3/4-inch outside door with 30 percent glass and an aluminum storm door, and all values are based on an inside temperature of 68° F. For your area degree-days, find the appropriate values in table 31, then calculate your expected savings in table C-36 (Appendix C). In addition to these savings, you will also reduce infiltration heat losses.

For example, if you live in an area with 5,000 degree-days and

TABLE 31
BTU Saved by Adding Storm Windows and Doors

Degree-days	*Add storm to:* Single-pane Window: Wood	Aluminum	Double-pane Window: Wood	Aluminum	Door
1,000	13,000	15,000	6,000	6,000	6,000
1,500	20,000	22,000	9,000	10,000	9,000
2,000	28,000	31,000	11,000	14,000	13,000
2,500	35,000	38,000	14,000	17,000	15,000
3,000	42,000	46,000	18,000	21,000	18,000
3,500	49,000	54,000	20,000	24,000	21,000
4,000	56,000	62,000	23,000	28,000	24,000
4,500	72,000	68,000	25,000	30,000	27,000
5,000	68,000	74,000	28,000	33,000	30,000
5,500	75,000	83,000	31,000	36,000	33,000
6,000	82,000	92,000	34,000	40,000	37,000
6,500	88,000	98,000	37,000	43,000	40,000
7,000	94,000	105,000	40,000	47,000	43,000
7,500	101,000	112,000	43,000	51,000	46,000
8,000	108,000	120,000	46,000	54,000	49,000
8,500	113,000	126,000	47,000	57,000	51,000
9,000	118,000	132,000	49,000	59,000	53,000
9,500	123,000	136,000	51,000	61,000	55,000
10,000	128,000	143,000	54,000	64,000	58,000

have wood casement windows, the addition of storm windows will save you 68,000 Btu per square foot of window. Heat-loss reduction through a window with an area of 10 square feet would be 680,000 Btu per year, which would mean a savings of about 7 gallons of oil, 10 Ccf of gas, and 200 kwh of electricity—or about $5 for oil, $3 for gas, and $10 for electricity. If you add a storm window to a double-glazed window, your savings would be about half these values.

Energy Savings: "Better Value" Calculate the expected heat loss with storm windows and doors added, then subtract that heat loss from your previously calculated heat loss (tables C-11 and C-12, Appendix C). If you plan to reduce the inside temperature, you must calculate the "Before" and "After" heat losses at the new inside temperature. The "Before" value in this case will be the "After" value calculated in table C-28 (Appendix C).

Calculate your expected savings in table C-37 (Appendix C).

Install Weather Stripping and Caulking

Weather-strip, caulk, and seal all cracks, openings, and gaps around your windows, doors, and walls. For windows and doors facing the wind, the best way to find large leaks is to move your hand around the edge of each window and door on a cold, windy day and feel for cold, entering air. For windows on the downwind side of the house, create a little dust in the air near window or door, and watch where and how fast this dust moves through the cracks. A fogged storm window indicates a leaky window on the downwind side of the house. Moist air leaking past the primary window condenses on the cold storm window; and the more condensation, the bigger the leak.

Many types of weather stripping are available on the market. Find the type that best suits your needs. Foam rubber or foam plastic weather stripping is easy to install and does a good job of sealing cracks. Felt stripping doesn't expand and contract well and, where stripping must mold to an irregular surface, is not as effective as rubber or plastic foam. You can buy special types of weather stripping for the sides and bottoms of doors.

Caulk any cracks that let air enter around window and door frames and at various other joints around the walls of your house. Some common sources of air leaks include areas around water pipes for outside faucets; where the house foundation meets the siding; where the porch meets the house; and where cracks exist in the masonry construction. If your house has wood, aluminum, or vinyl siding, be sure that all corners are well-sealed and that no gaps exist in the siding. Caulking is not a one-time job. Caulking material dries and breaks up after a few years, and then you need to recaulk. To keep recaulking to a minimum, don't be frugal when you buy caulking material. The expensive types of caulking (latex, butyl, or polyvinyl bases—or even more expensive but more durable silicones, polysulfides, and polyurethane-base caulkings) last longer than the cheaper oil-based or resin-based types.

Energy Savings The cost of weather stripping and caulking is small, but the savings can be considerable. Trying to calculate the exact savings is nearly impossible, however, because of the subjective nature of heat-loss calculation. To gain a rough idea of the energy saved by weather-stripping a window or door, find the leakage-reduction factor in table 32. These factors indicate the reduction of infiltrating air in cubic feet per hour per foot of crack sealed. To choose a value from the table, you must decide if your windows and doors are loose-fitting or have an average fit. A loose-fitting window or door is defined as one with a crack width of 3/32 of an inch or larger; and an average-fitting window or door is one with a crack width of 1/16 of an inch or less. All of your windows and doors may not fall into the same category, so either separate your windows and doors into one of the two categories or just assume that they all belong in one or the other category.

After choosing the appropriate leakage factor(s) in table 32, use table C-38 (Appendix C) to calculate the expected reduction of air infiltration and your energy savings. If you also plan to install storm windows or doors, include them in your calculations because they also help reduce infiltration.

Keep Your Furnace Serviced

An oil burner needs cleaning and adjusting at least once a year for proper operation. As a furnace gets dirty, its efficiency de-

TABLE 32
Factors for Reducing Infiltration by Weather-stripping

	Leakage Savings Factors (Cfh per foot of crack)	
	Window	*Door*
Loose-fitting windows or doors, no storm or weather stripping. Add:		
weather stripping only	17	26
weather stripping and storm	19	29
storm only	8	16
Loose-fitting windows or doors with storm, no weather stripping. Add:		
weather stripping	12	13
Average-fitting windows or doors, no storm or weather stripping. Add:		
weather stripping only	5	9
weather stripping and storm	6	12
storm only	3	6
Average-fitting windows or doors with storm but no weather stripping. Add:		
weather stripping	4	6

creases; and only a small decrease in efficiency results in a large increase of fuel use.

Insulate Your Furnace Boiler

If you have a hot-water heating system, wrap the furnace with 6 inches of fiberglass or mineral wool insulation. Use the type without paper backing, or remove the paper from the insulation. This action will reduce the standby heat loss from the boiler. Be sure you don't block air flow to the burner or cover up any controls for the furnace.

Keep the Fireplace Chimney Damper Closed When Not in Use

If you don't use your fireplace or only use it infrequently during winter, fabricate some kind of tight seal for the fireplace. A normal damper seal is poor at best; because the explicit purpose of the chimney is to draw air from the house, a significant amount of it flows through the damper seal—more than would be lost through an equal sized opening around a window or door.

Let the Sunshine In

During the day, open your curtains on the south side of the house. This does not decrease the heat loss, but it does increase heat input from the sun. If you are concerned that the sunlight will fade your furniture, install a sheer curtain to diffuse the direct sunlight. Or keep the curtains closed, but back them with a black liner that absorbs all the light rather than reflecting it out the window. Close all the curtains at night. The layer of air trapped between curtains and window helps to reduce conductive heat loss.

Cooperate with Nature

Plant evergreens and shrubs around the windy side of your house. They will block the wind and decrease heat loss from both conduction and convection.

Fill the House with Lots of People

Their body heat can replace some of the fuel needed to heat the house. But be sure to send them home at mealtimes or you'll spend more to feed them than you'll save on fuel.

Planning to Build a New House or to Remodel?

The best time to incorporate energy saving features in a house is during the building. Your primary concern is insulation. Ceiling insulation should have an R-value of 30 (9 inches of fiberglass, mineral wool, or the equivalent in blown insulation). The R-value for a typical wall (figure 14) is about 14, but you can increase this value to about 19 if you use 1-inch polystyrene sheeting rather than regular fiberboard sheathing or plywood. For the best possible wall insulation, use urea-formaldehyde foam instead of fiberglass insulation. This, in combination with polystyrene sheathing, gives an R-value of about 22. Use foam insulation around the window and door frames before you install the inside walls in order to reduce air infiltration into the house. Insulate your slab floor or crawl space as indicated in the previous sections. Don't expose any more of the basement wall to the outside than necessary, since heat loss through basement walls above ground level is considerably greater than through walls below ground. Storm windows and doors are a must. Single-unit storm windows are preferable to combination storm and screen units. Provide an independent air supply for a gas or oil furnace so the air for your furnace doesn't go through your house. For more details about how and where to insulate, one source to consult is your electric or gas utility company.

Obviously you must insulate as well as possible, but you can also incorporate other ideas to reduce heat losses and increase heat gains. A house should be rectangular to give the smallest wall area for a given floor area. (Actually a square house is best, but square houses are not as esthetically pleasing.) The two houses shown in figure 20 each have 1,500 ft^2 of floor area; but the rectangular house has 1,280 ft^2 of wall area, while the other house has 1,504 ft^2 of wall area.

Although a two-story house has half the attic area as a single-story house with the same living area, heat loss through the additional wall area generally equals that of the additional attic area in the single-story house. Heat loss through the basement, crawl space, or slab floor, however, is greater for a single-story house; so a two-story house will require less energy to heat than a single-story house with an equal living area.

The long sides of the house should face north and south, and

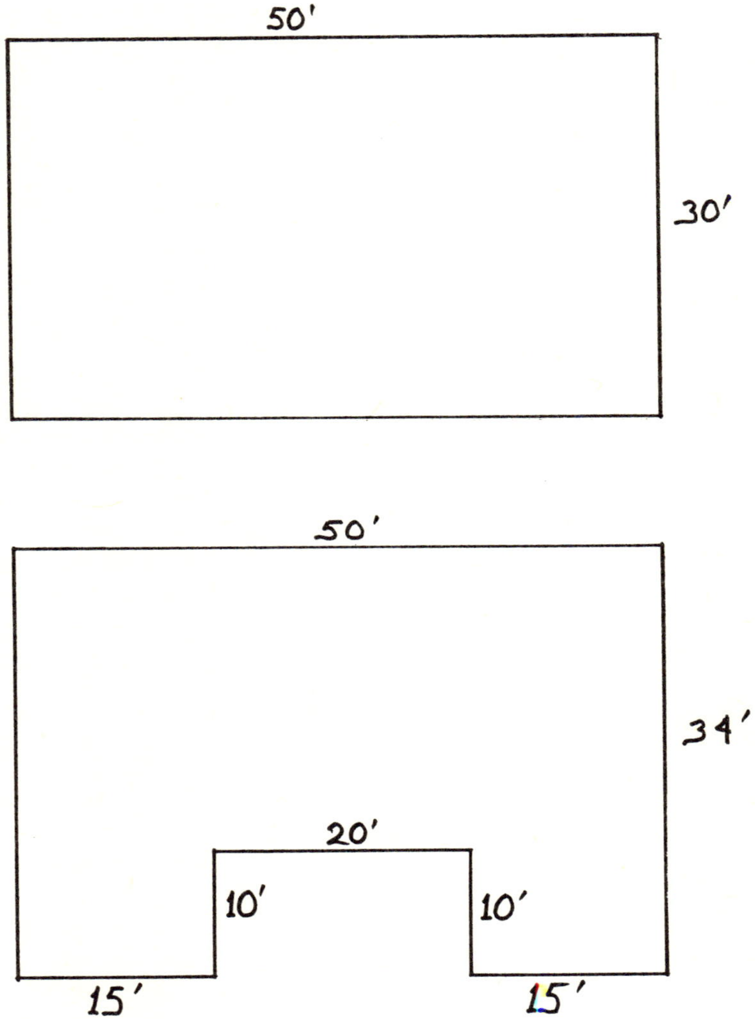

Fig. 20. Two houses with same floor area but different wall areas

windows on the north side should be smaller than those on the south side. In most regions of the United States, the coldest winds come from the north and west; and because windows lose more heat per area than any other surface, sizes of the north windows should be kept to a minimum. An important source of heat is the sun, so the largest windows should be on the south side. Since the south windows are generally sheltered from the wind, less conductive heat loss occurs through them than through the north windows. Most heat from the sun during the summer comes through the east and west windows, not the south windows; so large south windows will not provide large heat gain during the summer.

To help block the sun during the summer, extend your eaves about two feet beyond the house. These eaves will block the sun when it's high during the summer; but during the winter, when the sun is low, sunlight can enter the windows.

If you are planning to add an attached garage or enclosed porch, build on the windy side (probably the west end) of the house. If you plan to add both, build one on each end. These structures block wind during the winter and sunlight during the summer.

Fireplaces should have a glass screen in front that can be closed tightly to prevent the escape of room air. To provide sufficient air for the fire, install a duct from the outside of the house to the front of the fireplace floor; or install a grill behind the fireplace if the back of it faces the outside. A duct should have a damper at the outside wall to shut off the air when not needed. If you want your fireplace to supply more than a small fraction of its heat to the room, install a heat exchanger when the fireplace is built.

Although you may not be able to incorporate all of these ideas into a house, use them as guidelines. Books that go into considerably more detail about home construction and energy consumption are listed in the bibliography at the back of this book.

What kind of heating system should you install in your new house? A house should last from forty to sixty years. Although a heating system won't last as long as the house, changing from one type of heating system to another when replacing the old system is difficult and expensive. Therefore you should consider installing a heating system based on use of a fuel that will be available for many years. But what fuel will be plentiful in thirty, forty, or fifty years? Although nobody can be certain of the future, we can see indications of future trends and availability from the present situation.

Natural gas is probably the least plentiful fuel. In some areas of the United States, local gas utilities are not accepting any new customers, or accept only a limited number. Although natural gas may be the cheapest present fuel, it is in shortest supply.

Oil is a little more plentiful than natural gas, but world oil production is at, or almost at, its maximum; the amount of available oil will decrease in the near future—if not in the next five or ten years, certainly within the next thirty or forty.

Electricity is presently the most expensive fuel for most areas of the United States. Electric generating stations, however, can use two plentiful fuels to generate electricity that we can't use directly in our homes: coal and uranium. We should have an adequate future supply of electricity if the electric utility companies are able to construct enough coal and nuclear power plants. Thus if you want a conventional type of heating system, you should install electric heat, preferably a heat pump.

Although a properly operating heat pump uses about half the electricity of resistance heat, an entire nation of homes using heat pumps could become a serious problem for the electric utility companies. Heat-pump efficiency decreases at low outside temperatures, so the heat pump uses more electricity to produce the necessary heat for a house; and when the temperature is low, the make-up heaters turn on. During a cold spell, electric utility companies would face excessive demands for power. A partial solution to this problem is to install a heat pump that uses natural gas or oil for make-up heat. Some future systems may include heat-storage capacities to avoid the excessive load problem.

But the most plentiful energy for the next thirty years to five billion years is solar. Although the present cost of solar heating may be large, the fuel savings will be considerable over many years of operation. In addition, you won't have to worry about oil embargoes or gas shortages—just hope the cloud cover doesn't get too heavy! If you do install a solar heating system, make sure you employ a reputable contractor and designer to do the work.

4.

Hot-Water Heater

The energy used by the hot-water heater in most homes is second only to that used for home heating. With hot water readily available at every sink, we tend to forget about the energy needed to heat the water.

Hot-water heaters use oil, gas, or electricity to heat the water. Oil and gas hot-water heaters burn oil or gas under a tank of water; and an electric hot-water heater has one or two heating elements inside the tank that heat the water. Figure 21 shows several types of hot-water heaters.

The energy needed to heat the water depends on the difference between the temperature of the cold water entering the heater and the temperature of the hot water leaving the heater. As an example, let's calculate the energy cost required to heat 100 gallons of water if the cold water entering the heater has a temperature of 60° and the hot water leaving the heater has a temperature of 140° F. From the definition of the Btu (1 Btu heats 1 lb. of water 1° F.), 100 gallons of water (834 lb.) heated by 80° F. (140° − 60° F.) requires 66,720 Btu (834 × 80° F.). The cost of 66,720 Btu depends on the type of energy used for the heat and the cost of that energy. Table 33 shows the cost of 100 gallons of hot water

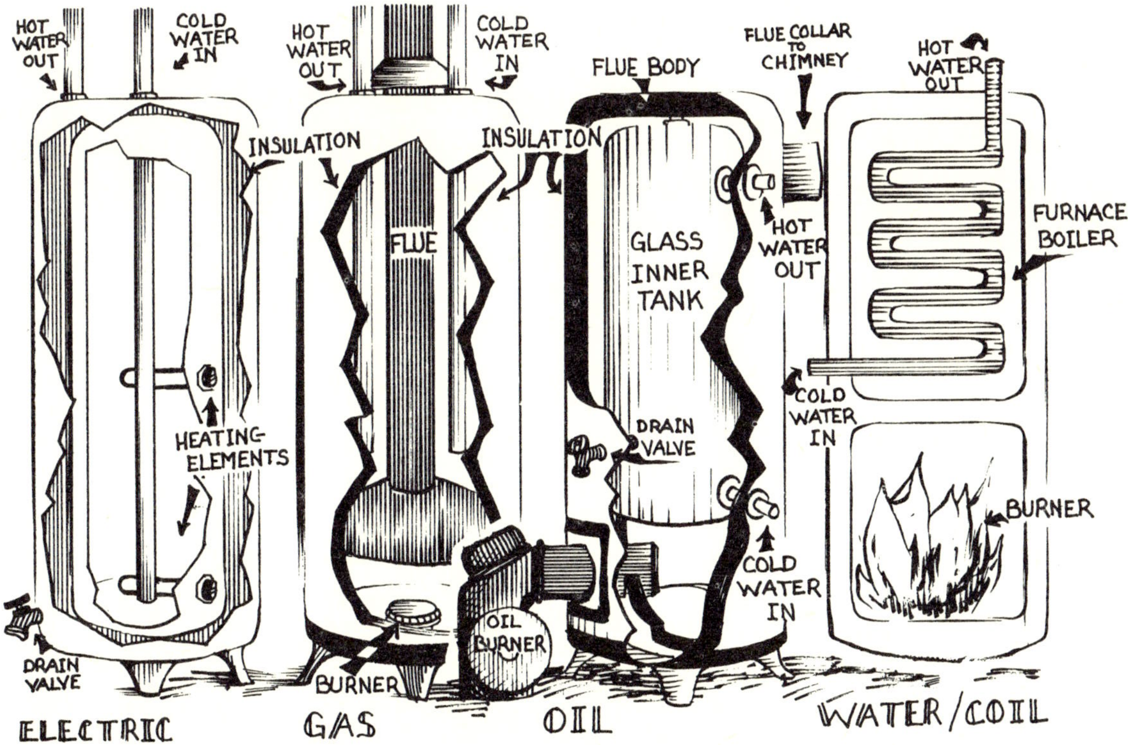

Fig. 21. Typical hot-water heaters

TABLE 33
Costs for Hot Water

Fuel and Costs		*Cost per 100 gallons of Hot Water*	*Cost per gallon of Hot Water*
Oil, per gallon	$0.75	$0.51	0.51¢
	0.70	0.48	0.48
	0.65	0.44	0.44
	0.60	0.41	0.41
	0.55	0.37	0.37
	0.50	0.34	0.34
	0.45	0.31	0.31
Gas, per Ccf	0.50	0.47	0.47
	0.45	0.43	0.43
	0.40	0.38	0.38
	0.35	0.33	0.33
	0.30	0.29	0.29
	0.25	0.24	0.24
	0.20	0.19	0.19
Electricity, per kwh	0.06	1.17	1.17
	0.05	0.97	0.97
	0.04	0.78	0.78
	0.03	0.59	0.59
	0.02	0.39	0.39

for the different types and costs of fuels. The numbers are based on an efficiency of 100 percent for electric heaters and 70 percent for gas or oil heaters. Note that for our example—fuel prices of 75 cents per gallon of oil and 35 cents per Ccf of gas—the cost of a gallon of hot water is about the same for both fuels (0.51 cents and 0.33 cents). But a gallon of water heated in an electric heater costs considerably more: 1.17 cents if the electricity costs 6 cents per kwh.

Your Costs for a Gallon of Hot Water

Now you must climb out of your easy chair again to obtain some information. You need to know the temperature of the water going into your hot-water heater and the temperature of the hot water coming out. To find the input water temperature, measure the cold-water temperature at a faucet near the hot-water heater (be sure to leave the faucet open awhile to flush out the water contained at room temperature in the inside pipes). The actual temperature for your cold water may vary depending on its source and the time of year. Water from wells remains at a fairly constant temperature all year—about 55° to 60° F.; but temperature of water from rivers and lakes can change from about 40° or 50° in winter to 60° or 70° F. in summer. If your cold-water temperature changes a lot during the year, you should measure throughout the year to find the average temperature or see if your water company keeps a record of average water temperature for your area.

Next you'll need to know the hot-water temperature. If your hot-water heater has a thermostat calibrated in degrees, take that temperature setting as the hot-water temperature. If the thermostat has no temperature scale or only a high, medium, or low setting, you must measure the actual hot-water temperature (and even if your thermostat is calibrated in degrees, you may want to check the scale by measuring the actual temperature).

Measure hot-water temperature at the faucet closest to the hot-water heater—or if the heater has a faucet at the output, measure the temperature there. Again be sure to leave the faucet open long enough to flush out the cooled water in the pipe from heater to faucet. When measuring the temperature of the water coming from the faucet, place the thermometer bulb in the running water—don't pour the water into a container and measure the temperature there, because the water cools quickly in a container.

If your thermometer scale doesn't extend up to the hot-water temperature, use a cooking thermometer. Or fill a container (preferably a quart or larger) half-full of cold water; place the thermometer in the water and read the temperature. Then turn on the hot water until it reaches its maximum temperature and fill up the container with hot water. Be sure the amounts of hot and cold water are equal. Read the final temperature of the water mixture. Now subtract the cold-water temperature from the final temperature, and add that difference to the final temperature to give the hot-water temperature. For example, if the half-filled container has cold water at 60° and the final temperature is 100° F. after the hot water is added, the difference is 40° F. Adding this difference to 100° gives a hot-water temperature of 140° F.

Now that you have the temperatures of hot and cold water, you can sit down and calculate costs. First subtract the cold-water temperature from the hot-water temperature to obtain the temperature rise. For 1 gallon of hot water, multiply the temperature rise by 8.34 to find the number of Btu needed to heat that gallon. Record the temperatures and do the calculations in table C-39 (Appendix C).

To find the cost of this energy, read the section below that pertains to your own type of hot-water heater: electric, gas or oil tank type, or gas or oil boiler type (part of your home heating system).

Electric Hot-water Heaters

Electric heaters are 100 percent efficient; that is, all of the electrical energy goes toward heating the hot water. Since some heat does escape through the walls of the heater, not all of the heat stays in the water (we will treat this standby heat loss as a separate expense later). To find the number of kilowatt-hours needed to heat 1 gallon of water, divide the Btu in table C-39 (Appendix C) by 3,413 Btu per kwh. Then, to find the cost of 1 gallon of hot water, multiply the kilowatt-hours needed to heat that gallon by the average cost per kwh. When considering the amount of money saved by conserving hot water, savings are calculated at the "last" cost per kwh. Calculate the cost of your hot water in table C-40 (Appendix C).

Gas or Oil Tank Type Hot-water Heater

First you must find the efficiency of your heater. For this you need two numbers, either from the unit itself or from the literature that came with the heater: the recovery rate (in gallons per hour at a given temperature rise) and the Btu input per hour. The recovery rate is a measure of the heat absorbed by the water. For example, if a hot-water heater has a recovery rate of 60 gallons per hour at a temperature rise of 100° F., the heat added to those 60 gallons of water is 50,040 Btu per hour (60 gallons per hour × 8.34 lb. per gallon × 100°F. temperature rise). This quantity is the useful energy. The Btu input is the actual energy needed to heat the water, so if the heater is rated at 75,000 Btu per hour input (the input energy), the efficiency (useful energy divided by input energy) would be 67 percent (50,040 Btu per hour ÷ 75,000 Btu per hour = 0.67 = 67 percent).

Because the efficiency is less than 100 percent, not all of the heat produced by the gas or oil ends up as heat in the water. The amount of heat actually absorbed by the water is the efficiency multiplied by the heat produced by the gas or oil. Or if we know the efficiency and the amount of heat that must be added to the water, the amount of heat which the gas or oil must produce is the amount of heat added to the water divided by the efficiency. In our previous example, we needed to add 66,720 Btu to 100 gallons of water at 60° in order to heat that water to 140° F.; so the amount of energy which the gas or oil must produce is 99,580 Btu (66,720 Btu ÷ 0.67 [efficiency]). Calculate the efficiency of your hot-water heater in table C-41 (Appendix C), then find the cost to heat a gallon of water.

Gas or Oil Boiler Type Hot-water Heater

The efficiency of this hot-water heater is exactly the efficiency of your furnace that you calculated in table C-7 (Appendix C); so use that value to calculate the cost per gallon of hot water in table C-41 (Appendix C).

To Shower or To Bathe: How Much Hot Water?

Of the total amount of hot water a family uses, the major quantity is for bathing, laundry, and washing dishes. Details of calculating the amount of hot water used by clothes washers and dishwashers will be given later in sections describing these appliances.

Most energy-saving tips state that a shower uses less hot water than a bath, but your own personal habits will dictate which practice actually uses less hot water. A short shower uses less hot water than a deep bath—but a long shower can use more hot water than a shallow bath.

How much hot water does your shower or bath use? To find out, you must know the flow rate (in gallons per minute) of your bathtub faucet and shower head. Again you'll need that quart or gallon container and a clock or watch with a second hand (a gallon container is preferable for measuring the flow rate of tub or shower because the flow rate is generally large and a quart container fills up too quickly). Station yourself next to the bathtub, turn on the hot water, place the container under the faucet, and measure the time it takes to fill. For a gallon container, divide 60 by the time taken (in seconds) to fill the container; the resulting number is the flow rate in gallons per minute (gpm). For a quart container, divide 15 by the time taken (in seconds) for it to fill. For example, the water from a faucet that fills a gallon container in 10 seconds has a flow rate of 6 gpm (60 ÷ 10 seconds = 6 gpm); and the water from another faucet that fills a quart container in 3 seconds has a flow rate of 5 gpm (15 ÷ 3 seconds = 5 gpm).

The task of filling a container with water may not sound difficult, but you may need a mop and pail if you aren't careful. The container should have a large opening on top so you don't have to worry about precise positioning under the stream. For the tub faucet, you can support the container on the bottom of the tub; but for the shower, you must hold the container high to get the full stream. A gallon of water weighs over 8 lb., remember, so make sure you have a good hold. If you think about the problem of holding the container, turning on the water, and looking at a watch, you'll realize that the entire operation requires about three hands and four eyes—so it's best to call for help; one person can do the timing while the other operates the faucet and container. The entire operation can be accomplished from outside the shower or tub (any different method is at the discretion of you and your partner).

Flow rates can be different for different faucets depending on the type of faucet and the length and size of the pipe supplying the faucet. The flow rate can also change with water pressure. For homes that obtain water from a commercial water company, pressure will probably be constant; but for a home with well and storage tank, pressure changes with the water level in the storage tank. If you have a well, you should measure the flow rates at the average pressure of the system, not at its highest or lowest pressure.

For your bathtub(s), record the times and calculate the flow rates in table C-42 (Appendix C).

The hot-water flow rate for shower heads is not as easy to determine as the tub flow rate. Some shower heads restrict the flow of water to give a good spray rather than just a large dribble, so the hot-water flow may not equal that from the tub faucet. And because the water used in a shower is a hot-and-cold mixture, you must measure some temperatures to find the hot-water flow rate.

Set the flow and temperature for your normal shower conditions, then measure the temperature and flow rate of the shower stream. Let's call these values the "warm-water readings." Then measure the temperatures of the hot water alone and the cold water alone. The fraction of the flow that is hot water is the difference between the warm-water and cold-water temperatures divided by the difference between hot-water and cold-water temperatures; that is, the warmer the water, the larger the hot-water fraction. If no cold water is used, the warm-water temperature is the hot-water temperature, and the hot-water fraction is 1. The flow rate of the hot water is the flow rate of the warm water multiplied by the hot-water fraction. If all this sounds confusing, consider the following sample calculation.

Suppose the cold-water temperature is 60°, the hot-water temperature is 140°, the warm-water temperature is 110° F., and the warm-water flow rate is 6 gpm. The temperature difference between the warm water and cold water is 50° F. (110° − 60°); and the temperature difference between the hot and cold water is 80° F. (140° − 60°); so the hot-water fraction is 0.625 (50° ÷ 80°). The hot-water flow is 3.75 gpm (0.625 × 6 gpm).

Measure the warm-water temperature and flow for each person in your family who takes a shower. Record these values in table C-43 (Appendix C), and calculate the hot-water flow for each.

Now back to the original question: Does your shower or bath use more hot water? To answer this, you must now measure the time required for the hot water to fill the tub to the desired level for a bath and the time required for a typical shower. Timing for the hot-water flow starts when you turn on the hot-water faucet, not when the hot water finally emerges from the faucet or shower head—because you want the total amount of time that the hot-water heater is supplying hot water. Once you have measured these times, fill in the blanks of table C-44 (Appendix C) for each member of the family, then calculate the amount of hot water used by each for a shower or bath.

For those members of the family who take either baths or showers, you can easily compare the amounts of hot water used for each. You can also compare hot-water amounts used by different members of the family.

Other Hot-water Uses

The amounts of hot water used by each person to shower and bathe remain about the same for that person unless he or she changes bathing habits from day to day. Dishwashers and clothes washers also use a definite amount of hot water for each use, and you can measure these daily hot-water amounts fairly accurately. But the amounts of hot water used for washing hands, washing dishes by hand, cooking, and other miscellaneous purposes do not remain constant. The amount of hot water used to wash dishes by hand, for example, depends on the number of dishes, pots, and pans to be washed, which can vary from day to day. The amount of hot water used to wash hands depends on the number of people at home and how often they need to scrub. Finding the exact amount of hot water used per day for these uses is difficult, but we can make some decent estimate. Although our estimate may not be exact, we know that hot-water amounts for these miscellaneous uses are considerably less than amounts used for bathing and laundry; and inaccuracies in the calculations for miscellaneous uses won't significantly affect our final calculation for total hot-water usage.

For each use listed in table C-45 (Appendix C) that is applicable to your home, estimate the amount of hot water used and the number of uses per day or per month. You can either roughly estimate or actually measure the total amount of water used and the hot-water fraction of that total amount. If you use only hot water to wash your hands, for example, the total amount of water you measure is the amount of hot water used. Always include the amount of water from the hot-water faucet that precedes the actual hot-water stream. If the hot water is too hot for washing hands and you add some cold water to cool it a little, estimate the fraction of the water that is hot water, then multiply that fraction by the total amount of water used. The resulting figure is the amount of hot water used. To be a little more exact, use the method outlined in table C-43 (Appendix C) for finding the amount of hot water used by a shower. If you wash dishes by hand, use table C-84 (Appendix C) in the section on dishwashers for calculating.

Even if you sit down and try to identify every use of hot water, you probably won't think of all of them. You may want to note whenever someone turns on the hot water for a week or two, or maybe tie a ribbon on each hot-water faucet to increase your awareness of hot-water use. As stated previously, the first step in energy conservation is simple awareness.

Standby Heat Loss

In addition to the energy needed to heat the water used in the home, some energy is required just to keep the water hot in the heater. The heater walls conduct heat from the hot water to the surrounding air, and this heat escape, called "standby heat loss," can be a significant part of the total energy used by the heater. A 50-gallon tank type heater can have a standby heat loss of 300,000 to 500,000 Btu per month, depending on insulation thickness and quality and temperatures of the hot water and room air. Standby heat loss for a boiler type system can be considerably higher.

To gain some idea of the cost of this heat loss, let's assume that a hot-water heater has a standby heat loss of 400,000 Btu per month. An electric heater, to make up this heat loss, would use 117 kwh (400,000 Btu ÷ 3,413 Btu per kwh) at a cost of $7.02 (at 6 cents per kwh). And, because they are only about 70 percent efficient, an oil or gas heater must produce 570,000 Btu (400,000 Btu ÷ 0.70 [efficiency]) per month to replace the 400,000 Btu lost. A gas heater would use 5.7 Ccf at a cost of $2.00 per month (at 35 cents per Ccf), and an oil heater would use 4.1 gallons of oil (570,000 Btu ÷ 140,000 Btu per gallon) at a cost of $3.00 per month (at 75 cents per gallon).

Again, each individual hot-water system is different, so you must find the standby heat loss for your own heater. If the literature that came with the heater gives the standby heat loss, you can calculate your cost per month and proceed to the next section. Most likely, however, there is no such information—the literature is lost, never existed, or was stolen by gremlins, so you need to do a few calculations. If you have a tank type heater, you can use the Approximation Method given below. If you want a more exact calculation, however, follow the instructions given in the On-time–Off-time or Vacation Method sections. These sections require much time and effort, so you may want to use the Approximation Method first, then when you have more time do the more exact calculations outlined in the later sections. If you have a furnace boiler system, you have no choice and must use either of the last two sections.

Approximation Method for Tank Type Heaters

To calculate the amount of heat loss, we can use the same method we used to calculate the conductive heat losses from a house. All we need is the area of the hot-water heater, the R-value of the heater wall, and the temperature difference between the hot water and room air. The R-value of a heater is approximately 3 if the heater has 1 inch of insulation, and approximately 5 if it has 2 inches of insulation. The areas of the different-size heaters are given in table 34. Measure the height and diameter (in inches) of your heater tank, then find the heater area (in square feet) in table 34 using the diameter and height values closest to your own values. For example, a tank with an 18-inch diameter and a 56-inch height has an area of 25.1 ft². Use the area of your tank and the approximate R-values mentioned to calculate the standby heat loss for your heater in table C-46 (Appendix C.)

On-time–Off-time Method

To replace the standby heat loss, the oil or gas burner or electric heating element turns on periodically, reheating the water. You can determine the amount of standby heat loss if you can determine when the burner or heating element turns on and off.

On a day when nobody will use hot water for several hours, record the times that the burner or heating element turns on and off. Some heaters may have long cycles—up to six or seven hours, depending on the amount of heat loss and the temperature setting at which the burner or heating element cycles on and off. To insure no use of hot water occurs, you may want to shut off the hot-water valve so that none will flow through the heater even if someone opens a hot-water faucet.

From the moment the burner or heating element turns off, measure the time until it turns on again. Call this time the "off-

TABLE 34
Areas of Tank Type Hot-Water Heaters (in square feet)

Height	*Diameter (inches)*												
(inches)	*14*	*15*	*16*	*17*	*18*	*19*	*20*	*21*	*22*	*23*	*24*	*25*	*26*
30	11.3	12.3	13.3	14.3	15.3	16.4	17.5	18.6	19.7	20.8	22.0	23.2	24.4
35	12.8	13.9	15.0	16.1	17.3	18.4	19.7	20.8	22.1	23.3	24.6	25.9	27.2
40	14.4	15.5	16.8	18.0	19.2	20.5	21.8	23.1	24.5	25.8	27.2	28.6	30.1
45	15.9	17.2	18.5	19.8	21.2	22.6	24.0	25.4	26.9	28.4	29.8	31.4	32.9
50	17.4	18.8	20.2	21.7	23.2	24.7	26.2	27.7	29.3	30.9	32.5	34.1	35.7
55	18.9	20.5	22.0	23.6	25.1	26.7	28.4	30.0	31.7	33.4	35.1	36.8	38.6
60	20.5	22.1	23.7	25.4	27.1	28.8	30.5	32.3	34.1	35.9	37.7	39.5	41.4
65	22.0	23.7	25.5	27.3	29.1	30.9	32.7	34.6	36.5	38.4	40.3	42.3	44.3
70	23.5	25.4	27.2	29.1	31.0	33.0	34.9	36.9	38.9	40.9	42.9	45.0	47.1

time." Then measure the time that the burner or heating element is operating, and call this the "on-time." Some burners don't make much noise when they turn on, so you must listen carefully; and electric heating elements make no sound at all, so watch for the indicator lights. If your electric heater has no lights that indicate when the heater is operating, this method is almost impossible to use.

Convert all times to hours and fractions of hours. Multiply the on-time by the number of Btu or watts input of the heater to find the amount of energy used during the on-time, then divide that amount by the total cycle time (on-time plus off-time) to obtain the amount of standby heat loss per hour. Although this calculation actually gives the amount of energy needed to replace heat loss rather than the amount of standby heat loss itself, the replacement energy is what you pay for—so that's what you really want to know. Record your times and calculate the amount of standby heat loss in table C-47 (Appendix C).

Vacation Method: Gas and Electric Only

This method utilizes your gas or electric meter. Its accuracy is probably not much better than the Approximation Method, but you can use it if you have trouble in timing just when your heater turns on and off.

The next time your entire family takes a vacation, read your gas or electric meter just before you leave and just after you return. The difference between these two readings is the amount of energy used by all the appliances that were operating, including the hot-water heater. If you have an electric hot-water heater, a weekend is enough time; but if you have a gas heater, you need to be gone for at least a week. In either case, you should choose a time when the home heating system will not be operating. In addition to the energy used by the hot-water heater, pilot lights for gas appliances and your refrigerator, freezer, clocks, and any lights left burning will also consume energy.

To separate the amount of gas used by the pilot lights, just subtract the pilot-light consumption that you calculated in table C-2 (Appendix C). To separate the amounts of electricity used by other appliances, you must take another few days away from home; but this time turn off the hot-water heater. When you do this, remember that you won't have hot water when you return—

so pick a time when you won't need it for a few hours after you return. Again, read your electric meter just before you leave and just after you return. The difference between these readings is the amount of energy used by your other appliances. With these heater-on and heater-off readings, calculate the amount of standby heat loss in table C-48 (Appendix C).

Pipe Heat Losses

Just as hot water in the heater loses heat through the heater walls, so hot water in the pipes also loses heat through the pipe walls—both when hot water is flowing in the pipe and when standing in the pipe after someone has used hot water.

While the hot water is flowing, heat loss is between 20 and 40 Btu per hour per foot of pipe. The exact value of the heat loss depends on the pipe size, the hot-water temperature, and temperature of the surrounding air. For hot water flowing through the pipe at a gallon per minute or faster, this heat loss in a 30- to 50-foot pipe is only 1 or 2 percent of the total heat in the water, and the water temperature drops only a degree or less. Because heat loss is small for pipes less than 50 feet long, we can ignore this amount.

When you turn off the hot water, however, water in the pipe quickly cools to room temperature, so the heat in that water is lost. Then the next time you turn on the hot-water faucet, the water from the hot-water heater must flush the cooled water from the pipe before the hot water reaches the faucet, and this amount can be significant. Although you can determine the amount of water in a pipe if you know the pipe size and length, this is not the total amount of water needed to flush out the pipe. The hot water from the heater pushes the cooled water ahead of it, but as it travels along the pipe the hot water loses heat to the pipe until the pipe is hot also. Therefore the wasted hot water is the total amount of water that flows from the faucet until water at the faucet reaches its maximum temperature.

To measure the amount of water needed to flush the cooled water from your pipes, place a container under the faucet and turn on the hot water until hot enough to use; then shut off the water and measure the amount in the container. Now, what is "hot enough to use"? You could measure the temperature of the water coming out of the faucet and say that 100°, 110°, 120° F., or any other temperature you choose is hot enough—but usually your fingers will decide the right temperature, so use them for this measurement.

Measure the amount of water needed to flush the pipe to each faucet in the house. Because different faucets in the house branch off one or two main hot-water pipes, allow time for the water to cool between each measurement—probably no more than 15 to 20 minutes—so that all the water in the pipe from heater to faucet has cooled before you turn on the next faucet. Record the number of gallons needed to flush out the pipes to each faucet in table C-49 (Appendix C). Even though these numbers represent a heat loss, much like the standby heat loss from the heater, this loss has been included in the previous calculations for the hot water used in your home. When you measure the hot water, the cooled water in the pipe is included in the measured amount; so the pipe heat loss represents part of the hot water used rather than an additional heat loss. And this pipe heat loss is a fixed loss—the only way to reduce it is to shorten your hot-water pipes, decrease the frequency of hot-water usage, or combine hot-water uses at one end of the house. You can decrease the frequency of usage to some extent, but you can't shorten the pipes without moving the water heater or the hot-water faucets.

Conservation Measures

The best way to save hot water is simply to use less. Each time you turn on the hot water, think! Do you really need that hot water? Do you really need that much?

Look at the uses listed in table C-45 (Appendix C). Start with the largest use and consider ways of reducing that amount significantly; then consider the next largest use, and so on. The small uses may be difficult to reduce. Some ways to help you reduce hot-water consumption and some ways to calculate your approximate dollar savings follow. You may be able to realize some additional hot-water savings from your dishwasher and clothes washer; these potential savings are listed in the sections concerning these appliances.

Reduce Hot Water for Showers and Baths

The amount of hot water used for a shower or bath depends only on the flow rate and the time the water is flowing—reduce either or both to reduce your hot-water consumption. Must your bath really be as full as you usually fill it? Do you really need to stand in the shower as long as you do? Do you need all that water tumbling out of the shower, or would less water do the job just as well?

Even though most showers have flow rates of 5 to 10 gpm, generally a flow of 3 to 4 gpm is adequate for a shower. To reduce your flow, you can take one of several actions. The easiest way to reduce the flow is simple—don't turn on the water full force. Adjust the faucet until you find the minimum flow that still gives an acceptable shower. If you don't want to bother with adjusting the hot-water flow, or if you have a single knob mixer control that doesn't allow control of the water flow, install a flow restrictor or shower flow control behind the shower head. See figure 22. These are inexpensive pieces of plumbing, easy to install, yet they effectively reduce the water flow. Or install a new shower head that restricts the flow. Most shower heads are not labeled with water flow rates, so finding one that effectively restricts the water flow may be difficult. Some companies, however, are now producing "energy-saving" shower heads with a flow rate of 3 to 4 gpm, yet give a good water spray.

Energy Savings Calculate the amount of hot water saved as a result of reducing your water flow, reducing your shower time (or bath-water fill time), or both. Do the calculations in table C-50 (Appendix C). Calculate the "after" hot-water flow rates and shower times or bath fill times in table C-50. Subtract the "after" value from the "before" value to find the number of hot-water gallons saved.

Fig. 22. Flow restrictor and shower flow control

If you now take a ten-minute shower with a hot-water flow of 4 gallons per minute, you are using 40 gallons of hot water for each shower. By reducing the hot-water flow to 2 gallons per minute and the shower time to five minutes, you would use only 10 gallons of hot water. The 30 gallons saved per shower would mean a savings of 15 cents per shower if you have an oil hot-water heater; 9 cents if you have a gas heater; and 29 cents if you have an electric heater. If you take one shower a day, your savings per month would be $4.50, $2.70, or $8.70, respectively.

Use Cold Water to Rinse Dishes

If you rinse your dishes before washing them, use cold water. Cold water will rinse as well as hot water, except for greasy foods. If you have a dishwasher that supposedly doesn't require the dishes to be rinsed, don't rinse them. If you find that some dishes aren't washed thoroughly, find which foods need to be rinsed off, then rinse only those dishes that need it.

Energy Savings Use table C-51 (Appendix C) to calculate the amounts of hot water and money saved.

Reduce Dish Rinsing

If you wash dishes by hand, experiment with different ways of rinsing the washed dishes, then find which way uses the least hot water. (See the section on dishwashers for more details.)

Energy Savings Use table C-52 (Appendix C) to calculate the amounts of hot water and money saved.

Repair Leaky Hot-water Faucets

A few drops per minute from a leaky hot-water faucet add up to millions of drops per year and hundreds of gallons of hot water wasted. To measure the amount of hot water wasted by a dripping hot-water faucet, place a cup measure under the dripping faucet and measure the water in the cup after one hour. If the cup runneth over, start over with a two-cup measure and plan to repair the leak as soon as possible. A one-cup-per-hour leak wastes 550 gallons of hot water per year.

Energy Savings Use table C-54 (Appendix C) to calculate the amounts of hot water and money saved.

Reduce Other Uses as Much as Possible

Consider each use and determine if it really needs hot water. If so, can you use less? When shaving, for example, fill a sink with hot water rather than use hot water running from the faucet.

Energy Savings Use table C-54 (Appendix C) to calculate the amounts of hot water and money saved.

Reduce the Hot-water Temperature

If the temperature of your hot water is higher than 140° F., set it at 140° F. or even lower if a lower temperature is satisfactory for your household uses. At this lower temperature, the hot-water heater has a smaller standby heat loss, the pipes lose less heat to the room, and the water requires less energy to heat. For appliances that use a definite amount of hot water, such as clothes washers and dishwashers,[1] the amount of energy saved is proportional to the temperature reduction. But for uses such as showers, baths, or hand washing when hot and cold water are mixed to produce a desired warm-water temperature, the amount of energy saved depends on one's personal habits. For example, if you regulate your shower temperature by turning the hot water on full force, then adjusting the cold-water flow to give the desired temperature, you will use less energy if you decrease the hot-water temperature. But if you don't turn the hot water on full force and instead adjust the flow of hot and cold water to give a desired flow and temperature, you will use the same amount of energy (if you keep the same water flow) regardless of the hot-water temperature.

Energy Savings The only way you can accurately calculate the amount of energy saved from a reduction of your hot-water temperature is to measure hot-water usage at the lower temperature and subtract that value from the previous usage value. This procedure requires you to recalculate the values in tables C-39 through C-45 (Appendix C), which repeats a lot of work. To keep this calculation somewhat simple, however, let's use a rough approximation: For each degree of temperature reduction, you save 0.3 percent of the total energy amount used for hot water. Using this approximation, calculate your expected savings in table C-55 (Appendix C). Thus if your monthly cost of hot water is $15, a temperature reduction of 10° F. would mean a monthly savings of 45 cents.

Turn Off the Hot Water When You're Away

When you take a vacation of more than a few days, turn off your hot-water heater and save the standby heat-loss energy. If you turn off the pilot light on your gas heater, be certain that it is properly operating again before you turn on the water.

Energy Savings The amount of energy saved is the amount needed to replace the standby heat loss for the period you are gone. The hot water won't be hot when you return, however, so you must subtract the amount of energy needed to reheat the water. Use table C-56 (Appendix C) to calculate the amounts of energy and money saved.

Keep the Hot-water Heater Properly Maintained

Drain a bucket of hot water from the bottom of your hot-water tank at least twice a year to remove sediment and mineral deposits that decrease the heater efficiency. Keep the burners of an oil heater clean, as efficiency can drop considerably. A drop in efficiency from 70 to 65 percent means an increase in hot-water cost of 7 percent.

Add Insulation to Your Hot-water Heater

If your heater has only 1 inch of insulation, an additional 3 inches can reduce the standby heat loss by more than half. Even if your heater has 2 inches of insulation, an additional 3 inches can save a lot. Use only fiberglass or mineral wool insulation without the paper backing. Or purchase a hot-water tank insulating kit.

Insulate Your Hot-water Pipes

This action will save only a little energy if any at all. The major heat loss from hot-water pipes is the result of hot water in the pipes cooling to room temperature. Hot water in an uninsulated pipe loses half its heat in ten to twenty minutes. By insulating the pipe, the time required for the hot water to lose half its heat increases 30 minutes to an hour, depending on water and air temperature and the size of the pipe. Insulation around the pipes may save you some energy if you use hot water at regular intervals (less than 15 to 20 minutes apart) during the day. If the times between hot-water uses are long, however, the hot water in insulated pipes will be just as cool as the water in uninsulated pipes, and you'll save no energy at all.

Planning to Build a New Home or Remodel?

Locate your hot-water faucets as close to the hot-water heater as possible. Design your house so that bathrooms, kitchen, and laundry room are grouped in the same part of the house as the hot-water heater. If some hot-water faucets are located far from the heater, calculate the monthly cost for the pipe heat loss; if that heat loss is more than the standby heat loss of a small heater, install a second heater near those faucets.

Install only shower controls that allow adjustment of the water flow. The best type is a single-knob mixer that allows control of the water flow. This type is better than separate hot and cold controls because once the proper mixture is set, you can adjust the flow without changing the water temperature.

Be selective when you purchase a hot-water heater. Calculate the efficiency of heaters you are considering, and find out the thickness of insulation. Consult *Consumer Reports* or other consumer publications that rate the different heaters by their operating costs. Don't buy a larger heater than you need—larger heaters have greater standby heat losses.

Install your heater in a warm place, not in an unheated basement, garage, or crawl space. Standby heat loss increases considerably when air temperature around the heater is below normal room temperature.

Install a heat trap in the hot-water pipe coming from your hot-water heater. A heat trap is simply a curve in the pipe, just like a trap below a sink, that blocks hot-water convection up the pipe.

5.

Kitchen Appliances

The amount of energy involved in the preparation of food can be a considerable part of a family's total energy budget. The entire process of food preparation includes the storage of food and clean-up afterwards, as well as actual cooking. All of these functions—storage, cooking, and clean-up—require energy, so let's consider the appliances we use for food preparation in the order of their energy consumption.

Refrigerator and Freezer

The refrigeration unit in a refrigerator or freezer is just a heat pump that removes heat from the inside of the unit (see chapter 3 for the operation of a heat pump). The cold coils are inside the refrigerator or freezer and the hot coils are outside, either on the back of the unit, under the unit, or on the sides or top just below the outside covering. Figure 23 shows the arrangement of coils for both manual-defrost and frost-free models.

In a manual-defrost freezer, the moisture in the air condenses onto the cold coils and turns to ice or frost. As the frost becomes thicker, it not only takes up space in the freezer that should be used to store food but also acts as an insulator between the

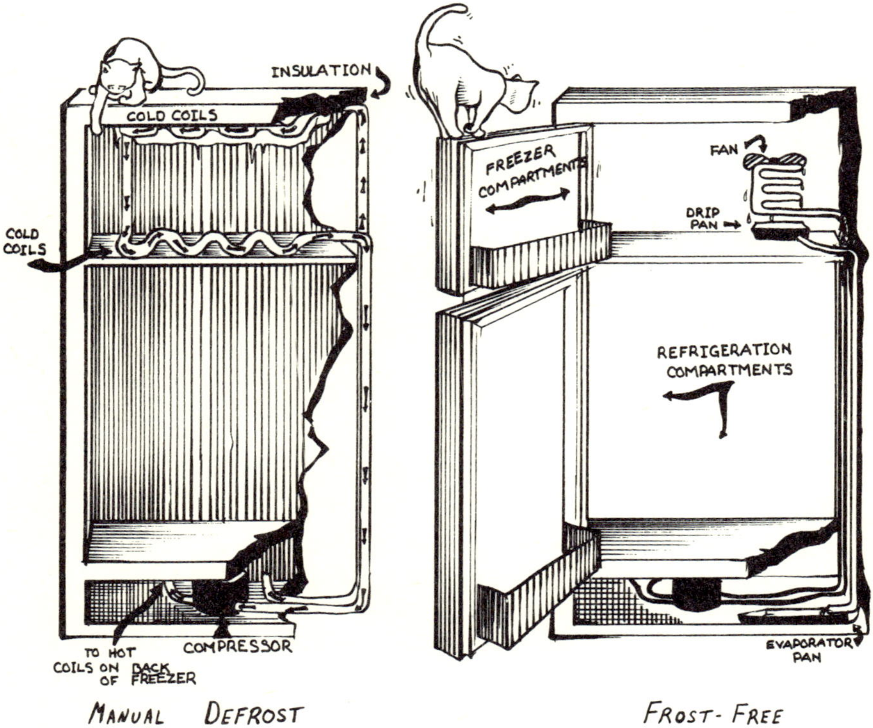

Fig. 23. Typical refrigerators

freezer space and cold coils. The heat that enters the freezer can't easily move through the frost to the coils, so the compressor must operate longer and use more energy than if the frost were not present. Therefore you should defrost a manual-defrost freezer frequently, not only to keep the freezer space clear but also to keep energy use at a minimum.

The cold coils for a frost-free freezer are not located in the freezer compartment, but behind or under the compartment. A fan circulates air through the freezer and over the coils. Water vapor in the air also condenses onto these coils, and the frost would soon build up on them if nothing else were to happen. Periodically (about twice or three times a day), however, the coils are heated, the frost melts, and the resulting water flows to a pan outside the freezer and evaporates. Extra energy is required both to heat the coils and to recool them, so most frost-free freezers use more electrical energy than the same freezers with manual-defrost operation—generally about 40 to 50 percent more.

But not all frost-free models use more energy than comparable manual-defrost models. The energy used by a refrigerator or freezer depends on the insulation thickness and efficiency of the refrigeration unit; and these can vary greatly from model to model. Some manual-defrost models use more energy than the same size frost-free model—thus we can't assume that all frost-free models always use more energy than all manual-defrost models (see the next section on "Relative Efficiencies and Costs").

The outside of the freezer door on some freezers and refrigerators becomes cold enough to condense water on warm, humid days. Some manufacturers have installed heaters around the doors to prevent such condensation. These antisweat heaters not only use from 10 to 30 kwh per month to operate, but they also produce heat that is conducted into the freezer—which means that the refrigeration unit must use even more energy. Some models have a switch that turns off the heater when not needed; but other models have no control switch, and the heater is operating constantly.

Relative Efficiencies and Costs

We can't easily define the useful work done by the refrigerator or freezer in terms of Btu removed from the unit or pounds of food preserved—so we can't really define an efficiency. All refrigerators and freezers, however, do the same job—keep food cold—so the most efficient model is the one that simply uses the least energy to maintain the desired temperature inside the refrigerator or freezer. The energy needed to keep the refrigerator or freezer cold depends on two factors: thickness and quality of insulation, and efficiency of the refrigeration unit itself.

But how do you determine the amount of energy used by your refrigerator or freezer? To find the amount used by most models, consult the *Directory of Certified Refrigerators and Freezers,* published semiannually by the Association of Home and Appliance Manufactures (AHAM).[1] This directory lists all refrigerator and freezer models of manufacturers that participate in the AHAM certification program; every model certified has an AHAM seal affixed. Presently, however, not all models list the energy use on the unit itself—you must find that information in the *Directory*.

The energy used by the different models can vary as much as 50% between similar sizes and types. For upright freezers with 15 to 15.9 ft^3 of storage space, for example, frost-free models use between 99 and 169 kwh per month, while manual-defrost models use between 68 and 100 kwh per month. You can see that the least efficient manual-defrost freezer (100 kwh) uses more energy than the most efficient frost-free model (99 kwh); and the least efficient frost-free freezer (169 kwh) uses more than twice the energy of the most efficient manual-defrost model (68 kwh). Therefore you must know the electrical consumption values of the different models in order to buy the most efficient one.

But what are "good" and "bad" values? Table 35 lists the monthly electrical consumption of different sizes and types of refrigerators and freezers. From the values in this table, you can observe some "cold facts" about the energy consumption of these appliances. Manual-defrost models generally use less energy than the same size frost-free models, and the larger units use more energy than the smaller ones, although some small units can also have large energy consumption values. Side-by-side refrigerators use more energy than the same size top-freezer models. One final fact not listed in table 35 is that a refrigerator with an ice and water dispenser on the front of the unit uses between 15 and 25 percent more energy than the same unit without the dispenser. If your family uses a lot of ice and cold water from a refrigerator with no ice and water dispenser on the door, however, the energy loss from opening and closing the door can increase the electrical consumption considerably. The unit with the ice and water dispenser may consume less energy during summer than one with no dispenser.

The AHAM *Directory of Refrigerators and Freezers* consists of about twelve pages of numbers—model codes, shelf areas, cubic feet of space, and kwh per month energy consumption—lots and lots of numbers. As you look through these numbers, however, you'll notice that many of them are the same. Of all the companies listed in the *Directory*, only a few actually manufacture their own units; the others purchase units from these manufacturers and attach their own labels. So these many pages of numbers condense to only a page or two of really different numbers. Your huge choice of many different models is really not so large as you thought.

Some of the energy ranges listed in table 35 aren't the actual ranges, but the only two energy values available for that size. For example, the 15-ft^3 partial-defrost refrigerators have a range of 51 to 80 kwh per month; but of all the models in this size, each has only one or the other of these two values—there are no values between. And some sizes have only one value—that is, all models of the size (sold under several different names) have the same consumption value.

Now let's look at a few more numbers in table 35. For the 12-cubic-foot partial-defrost refrigerators, the low value for this size (48 kwh) is also the lowest value for all of the partial-defrost refrigerators; and similarly, the model with the 84-kwh value in the 16-ft^3 frost-free refrigerators has the lowest value for all frost-free models. These low values are from the energy-efficient models that a few manufacturers are now starting to produce. Energy-efficient models have additional insulation and a better refrigeration unit, so they use less energy than a "conventional" refrigerator of the same size. Even though the cost of an energy-efficient model could be higher than a comparable conventional model, a purchaser will regain the extra cost in a few years from the energy savings.

To show what these energy consumption values mean in terms of dollars, table 36 shows the cost of the different energy consumptions for one year and for the expected fifteen-year lifetime of a refrigerator or freezer. If you purchase a unit that uses 140 kwh rather than one that uses 160 kwh, your saving is $12 per year, or $180 over the expected lifetime of the unit. Or suppose you are planning to buy a side-by-side refrigerator; the difference between the most and least efficient models is 49 kwh for the 18-cubic-foot models, while the difference between operating costs of these two models is $30 per year, or $450 over fifteen years. These savings are based on a "last" rate of 5 cents per kwh.

TABLE 35
Ranges of Energy Use for Refrigerators and Freezers (kwh per Month)

Refrigerators:

Size (Total Cubic Feet)	*Single-door Manual defrost*	*Top-freezer Partial-defrost*[a]	*Top-freezer Frost-free*	*Side-by-side Frost-free*
6.9 or less	31–68			
7.0–7.9	40	60	—	—
8.0–8.9	47	—	—	—
9.0–9.9	45–46	—	—	—
10.0–10.9	49–59	60	—	—
11.0–11.9	47–53	73–75	—	—
12.0–12.9	45–59	48–105	114–115	—
13.0–13.9	42–52	87–109	94–114	—
14.0–14.9	73	51–102	87–128	—
15.0–15.9	—	51–80	93–146	—
16.0–16.9	—	—	84–114	158
17.0–17.9	—	79	89–158	147
18.0–18.9	—	—	101–158	116–165
19.0–19.9	—	—	120–142	135–165
20.0–20.9	—	—	114–152	121–159
21.0–21.9	—	—	140	132–175
22.0–22.9	—	—	140	145–158
23.0–23.9	—	—	123	155–175
24.0–24.9	—	—	—	151–187
25.0–25.9	—	—	—	156–190
26.0–26.9	—	—	—	—
27.0–27.9	—	—	—	185

How Much Energy Does Your Refrigerator and Freezer Use?

If your refrigerator or freezer is new and you have the AHAM *Directory,* look up the value for its monthly energy consumption. However, if your refrigerator or freezer is more than two or three years old or is not listed in the *Directory,* you must measure the energy consumption yourself. The method you will use depends on the type of refrigerator or freezer you own and how much information you have about the unit. If you have a manual-defrost model, can easily hear when it turns on and off, and are willing to spend from 20 minutes to a few hours listening for it to do so, use the on-time—off-time method. Or use the electric meter method,

TABLE 35 (*Continued*)

Freezers:

Size (Total Cubic Feet)	*Upright Manual-defrost*	*Upright Frost-free*	*Chest Manual-defrost*
6.9 or less	62–68	—	37–69
7.0–7.9	—	—	37–59
8.0–8.9	—	—	65–68
9.0–9.9	72	—	70–78
10.0–10.9	75	—	69–70
11.0–11.9	90–98	—	76
12.0–12.9	79–90	—	85
13.0–13.9	60–106	128	—
14.0–14.9	81–99	—	98
15.0–15.9	68–100	99–169	82–104
16.0–16.9	88–136	116–137	—
17.0–17.9	75	129–153	—
18.0–18.9	79–116	143	100
19.0–19.9	94–99	152–160	124
20.0–20.9	83–122	149	100–118
21.0–21.9	94–102	199	—
22.0–22.9	—	—	—
23.0–23.9	131	—	105–138
24.0–24.9	—	—	123
25.0–25.9	—	—	114–141
26.0–26.9	—	—	114
27.0–27.9	120	—	138
28.0–28.9	—	—	160
29.0–29.9	—	—	—
30.0–30.9	—	245	—
31.0–31.9	170	—	—

Note: all values listed are for units with antisweat heaters off if a switch is present. If the antisweat heater switch is turned on, the kilowatt-hour consumption is greater.

[a] A partial-defrost refrigerator has no frost-free freezer section. These units are usually called "automatic defrost refrigerators."

Source: AHAM *Directory of Certified Refrigerators and Freezers*, June 1977.

especially if you have a frost-free unit. These methods are described in the following sections.

AHAM Directory The easiest way to find the amount of energy used by your refrigerator or freezer is to look up the value in the AHAM *Directory of Certified Refrigerators and Freezers*. The values listed in this directory are for "average" conditions—that is, for average room temperatures; they take into account the energy needed both for cooling food that you put into the unit and

TABLE 36
Cost of Different Energy Uses per Month

Kwh per Month	*Cost per year at Different Electrical Rates*			*Cost per Fifteen Years at Different Electrical Rates*		
	$.04/kwh	.05/kwh	.06/kwh	$.04/kwh	.05/kwh	.06/kwh
20	$ 10	$ 12	$ 14	$ 144	$ 180	$ 216
30	14	18	22	216	270	324
40	19	24	28	288	360	432
60	29	36	44	432	540	648
80	38	48	58	576	720	864
100	48	60	72	720	900	1,080
120	58	72	86	864	1,080	1,300
140	67	84	101	1,010	1,260	1,510
160	77	96	115	1,150	1,440	1,730
180	86	108	130	1,300	1,620	1,940
200	96	120	144	1,440	1,800	2,160
220	106	132	158	1,580	1,980	2,380
240	115	144	173	1,730	2,160	2,590

cooling the air that enters the unit when you open the door. The amount of electricity used by your unit may differ by 20 to 30 percent from the values listed because of different inside or outside temperatures, greater or lesser use of the refrigerator or freezer, and because amounts of energy used by different units of the same model can differ.

Your unit should be listed in the AHAM *Directory* if you purchased it within the last two or three years. Most models of appliances are on the market for several years, not changed annually like car models. If you can't find your own model listed but find one of the same size, differing from your model number by only a single number or letter, the new model is probably the same unit as your model except for a different shelf arrangement or other internal change. The basic wall construction, cooling unit, and energy consumption should be the same. List the energy consumption of your refrigerator, freezer, or both in table C-57 (Appendix C).

On-time–Off-time Method This is the same method as used to find the standby heat loss from a hot-water heater; similarly, it can tell you the amount of standby heat gain of your refrigerator or freezer. A refrigeration unit cycles on and off periodically to maintain the desired inside temperature. The duration of one cycle begins when the unit turns on once and ends when it turns on a second time. You must measure both the time the unit is on and the time required for one cycle.

To measure these times, you must keep the refrigerator closed for at least one cycle, which may last from twenty minutes to an hour or longer depending on the particular characteristics of your unit. After the door is closed for the final time after a meal, wait 15 to 20 minutes for the inside temperature to reach an equilibrium. Wait at least an hour if you have placed warm food in the unit. Record when the unit turns on, the time it turns off, and the time it turns on again in table C-58 (Appendix C); then calculate the energy use per month. For example, if your refrigerator turns on at 2:30:00, turns off at 2:55:30, and turns on again at 3:34:15, the "on-time" is 25.5 minutes (2:55:30 − 2:30:30 = 0:25:30 = 25.5), and the cycle time is 64.25 minutes (3:34:15 − 2:30:00). The fraction of the time the refrigerator is operating is 25.5 minutes ÷ 64.25 minutes = 0.40. Amount of monthly heat gain is the product of the wattage, the fraction of the time the refrigerator is on, and the average number of hours in a month. If your refrigerator uses 250 watts when operating, its monthly heat gain would be 250 watts × 0.40 × 732 hours per month ÷ 1000 = 73 kwh.

If the room air temperature changes greatly from summer to winter, you should do the calculation in both seasons to find summer consumption and winter consumption.

During normal operation, however, the unit door doesn't remain closed. Each time it opens, air enters the refrigerator or freezer and warms the inside air—and, of course, any warm food placed into the unit also warms the inside air. To remove this extra heat, a refrigerator uses about 20 percent additional energy over the heat gain through the unit walls; a freezer uses an additional 10 percent. So to find the total amount of energy used by both refrigerator and freezer, multiply the heat gain of your refrigerator by 1.2 and the heat gain of your freezer by 1.1. These factors are included in the calculation in table C-58 (Appendix C). For example, if the heat gain of your refrigerator is 73 kwh per month, the total amount of energy it uses per month would be 73 kwh × 1.2 = 88 kwh.

Electric Meter Method This method uses your electric meter to find the energy used by your refrigerator or freezer during a period of several hours or a couple of days. To prevent the energy used by other appliances from interfering with the calculation, select a time when no other appliances will be operating. Turn off as many appliances as possible—especially the electric hot-water heater—and any other appliance for which amount of energy use is difficult to calculate. Because all home heating systems use electrical energy for some part of their operation and because consumption of this energy is hard to calculate, you should do this present calculation only when house heat is off. You need at least 6 to 8 hours for manual-defrost models and at least 36 to 48 hours (a weekend away from home) for frost-free models.

Read your electric meter before and after the selected period. For short periods (less than one day), you should read your meter to the nearest tenth of a kilowatt hour (see Appendix A for details of interpolating). The difference between these readings is the electrical energy used by your refrigerator and freezer, plus the energy used by any other appliances also operating during this time. Calculate the energy used by the other appliances and subtract that energy from the total energy used during the entire time. The remaining energy figure is the amount used by your refrigerator or by refrigerator and freezer, if you have both. If you own both refrigerator and freezer, you can only find the amount of energy each consumes by using another method to find the amount consumed by one. If one unit is a manual-defrost model, you can use the on-time–off-time method to determine the energy used by that unit; the rest of the energy is the amount used by the frost-free unit. For example, suppose you are gone for five days (120 hours); your initial meter reading is 7,583, and your final reading is 7,615. The heat gain of your refrigerator and freezer (assuming you have both) plus any lights left burning is 32 kwh (7,615 − 7,583). If you have a 60-watt light burning for 7 hours per day, it will use 2 kwh during the five-day period. Therefore the heat gain of the refrigerator and freezer is 30 kwh per 120 hours, or 0.25 kwh per hour (30 kwh ÷ 120 hours). The monthly heat gain of both units is the heat gain multiplied by the average number of hours in a month, or 0.25 kwh × 732 hours = 183 kwh.

Keep the refrigerator or freezer door closed during the selected time. The energy thus calculated is the amount needed to remove the conducted heat gain into the unit. It does not include the amount of energy needed to cool food placed inside the refrigerator or freezer nor does it include the energy needed to cool the air entering the unit whenever the door is opened. To remove this extra heat, a refrigerator uses about 20 percent additional energy over the amount required to remove the heat gain of the unit; a freezer uses about 10 percent additional energy.

Record your initial and final meter readings in table C-59 (Appendix C), then do the calculations for the energy use.

Conservation Measures

You can do very little to save energy with your present refrigerator or freezer. The two properties that determine the cost of operation are the efficiency of the cooling system and the quality and thickness of the insulation—two factors you can't change. You can prevent the cost of operation from increasing, however, if you keep the cooling system operating at its maximum efficiency and keep the temperature around the refrigerator or freezer as low as possible.

During the heating season, your actual savings will be less than calculated here, especially if you have electric heat. Refer to chapter 9 for details.

Keep the Cooling Coils Clean Dust on the hot coils and frost on the cold coils insulate them, retard heat transfer, and therefore decrease efficiency of the cooling system. Defrost manual units when the frost is over 1/4-inch thick. Keep the hot coils clean by vacuuming them twice a year or more, and don't block the air flow around the hot coils.

Cover All Liquids in the Refrigerator and Freezer Vapor from the liquids condenses on the cooling coils, causing more frost to form. In a frost-free refrigerator or freezer, food can also dry out quickly.

Place the Refrigerator and Freezer in a Cool Place Keep your refrigerator or freezer out of direct sunlight and away from heat radiators or convectors. A freezer used for long-term storage and opened infrequently can be placed in an unheated basement or garage.

Energy Savings A refrigerator uses approximately 2 percent less energy for each degree of room temperature decrease, and a freezer uses 1.5 percent less energy. If you move your unit to a cooler place or decrease the room temperature, find the new average room temperature and calculate your expected energy savings. For example, if you move your freezer from an upstairs room to the basement where the temperature is about 5° F. less than the upstairs temperature, you can save about 7.5 percent of the energy (5° F. × 1.5 percent). If your freezer uses 120 kwh per month, the savings would be about 9 kwh, or about 45 cents, per month. Your total savings over a fifteen-year lifetime of the freezer could be as much as $80.

If the room temperature changes greatly during the year, calculate the energy savings for the different temperatures. Note that energy use will increase if the new room temperature is higher than present room temperature during the summer months. Calculate your expected savings in table C-60.

Don't Set the Temperature Too Low in the Refrigerator or Freezer A refrigerator should have a temperature of 37°, and a freezer's temperature should be 0° F. for long-term storage. If you have both refrigerator and freezer, use the freezer for long-term storage and the refrigerator freezer for short-term storage (less than a month); the refrigerator freezer then can have a temperature of 10° to 15° F.

Keep the Antisweat Heater Off If your refrigerator or freezer has an antisweat heater with an on-and-off switch, be sure the heater is off when not needed. Turning off a heater that uses 20 kwh per month means a saving of $1 per month.

Be Sure the Gasket Around the Unit Door Seals Properly. Close the door on a piece of paper; if the paper slips out easily, the gasket is worn and you should replace it.

Don't Open the Unit Door More Than Necessary Remove everything for a meal at one time—don't open the door for one item at a time. And don't stand in front of an open unit door trying to decide what you want or you'll end up with both higher electric bill and cold feet—neither of which is desirable.

Purchasing a New Refrigerator or Freezer? Unfortunately many refrigerators and freezers are purchased under pressure. If your old unit suddenly gives its last gasp, you may buy a model about the right size that can be delivered as soon as possible; but later you may find that you bought one that costs a lot to operate. If your old unit suddenly quits, rent another one, use your neighbor's, or even eat out until you can be sure you're buying the most efficient model available. The extra cost and effort may be a small fraction of the money later saved on electricity. If indications exist that your present freezer or refrigerator is about to go, do some investigating *before* its last gasp so that you're prepared to buy a new one when the time comes.

First decide what size and type you want. Don't buy a larger unit than you need—the larger models usually consume more energy than the smaller ones. Although all energy-saving tips repeat that a frost-free model costs from 35 to 50 percent more to operate than a manual-defrost, these are only average values; some manual-defrost models use more energy than a same-sized frost-free model. Be sure to check the energy consumption figures.

Consider an energy-efficient frost-free model if you don't want a manual-defrost or if manual-defrost models are too small. If the price frightens you, calculate the fifteen-year operating cost for that model and the operating cost for a model you think you can afford—now see which frightens you more! Avoid side-by-side refrigerators and refrigerators with ice dispensers on the front; they use more energy. With a little effort, you should be able to find a satisfactory refrigerator or freezer that uses less than 100 kwh per month.

Energy Savings To gain some idea of how much energy and money you can save with the purchase of a more efficient refrigerator or freezer, subtract the energy consumption of a new model from the energy consumption of your present model. Do the calculation in table C-61 (Appendix C), then find your expected savings.

Cooking Appliances

The process of cooking is just the heating of food, but cooking food is a lot like using a sledge hammer to crack a peanut. The energy used to heat an oven is much more than that actually needed to cook the food. Most of the heat generated by a conventional oven, cook top, or grill is lost to the room air by conduction through the oven walls or by direct convection or radiation from a cook top, or grill surface. Only a microwave oven can direct a large fraction of the energy to where it is needed—to the food, not the kitchen air.

Of all the energy used in the home, the most difficult amount to measure is that used for cooking. A kitchen can contain a multitude of appliances, each for a special function—toaster, grill, waffle iron, crockpot, and so on. The wattage of each appliance differs from model to model, and the time they operate varies from use to use. Therefore the energy used by a particular appliance can vary greatly from day to day, week to week, and month to month; so to measure the amount of energy consumed in the kitchen, you would need to record all the energy used by each appliance.

To find the energy needed to cook a roast, bake a cake, or fry a chicken, you must know the wattage of the appliance (or the rate of gas consumption for gas ovens and cook tops) and the operating time of the appliance. Appliances such as toasters, broilers, or blenders use energy continuously, so their simple operating time is easy to measure. But other appliances such as ovens, electric grills, and waffle irons have thermostats that periodically switch the electricity or gas on and off to maintain the desired temperature; for these appliances you must find the actual on-time of the gas or electricity.

The problem with calculating the amount of energy used for cooking is that a family doesn't have a monthly, weekly, or even daily pattern of appliance use in the kitchen. Each day the meals are different, so a "typical" or "average" day doesn't exist; the only way to obtain an accurate accounting of energy consumed is to measure the amount of energy used to prepare each meal—and that would be a long, tedious task. The following sections present some methods, however, that allow you to calculate the approximate energy use without requiring you to stand over a hot stove with watch, paper, pencil, calculator, and hungry children at your feet wanting their dinner. (Just try explaining to them the importance of knowing the number of kilowatt hours needed to prepare a meal when all they want is the meal!)

Ovens

Ovens come in all shapes and sizes and with different wattages or gas consumption rates. To calculate the energy used by your oven, you must know the wattage of the heating element(s) in an electric oven or the consumption in cubic feet per minute of a gas oven and the time the oven is operating. The instructions for your oven may list the wattage or gas consumption rates, or the oven may have a small plate that lists this information. If the information is not available, you can measure these values yourself. To calculate the energy needed to preheat and maintain the desired temperature in your oven, read either of the following sections for ovens, whichever applies.

Conventional Electric Oven A small electric oven has only one heating element for both baking and broiling, but a large oven usually has a broil element on top and a bake element on the bottom. In most ovens with a bake and broil element, the broil element turns on for part or all of the preheat cycle and only then does the bake element operate periodically to maintain the desired oven temperature.

If the wattages of your broil and bake heating elements are listed in the instructions or on the oven, proceed to table C-62 (Appendix C) and record those values in the appropriate blank. If they are not listed (they should be listed separately because you need the wattage of broil and bake elements separately), determine the wattage of each using the rotating disc method (see chapter 2). Record the data in table C-62 (Appendix C). If both broil and bake elements turn on during the preheat cycle, you must find the wattage for both together as well as for each element alone. Wattage of the bake element plus wattage of the broil element should, if all calculations are correct, equal the wattage of the two when on together.

Now you must time the durations that bake and broil elements are operating. Measure the times during actual baking conditions; prepare some food that bakes at 350° F., and use table C-63 (Appendix C) to record the times that the bake and broil elements turn on and off during the preheat cycle. For example, suppose you turn on your oven at 3:00:00; the broil element turns off at 3:02:45, and the bake element turns off at 3:11:25. Thus the broil element was on for 2 minutes 45 seconds, and the bake element was on for 11 minutes 25 seconds. Or, converting these times to minutes and fractions of minutes (divide the seconds by 60), the broil element was on for 2.75 minutes, and the bake element was on for 11.42 minutes.

After the oven is preheated, estimate the fraction of time that the bake element is operating while the food is baking; or you can measure the on-times and off-times for the bake element, then calculate in table C-63 (Appendix C) the fraction of operating time. For example, suppose the bake element is on for 1 minute 30 seconds (1.5 minutes) and off for 1 minute 10 seconds (1.17

minutes); the cycle time is 2.67 minutes (1.5 + 1.17). The fraction of the time the element is on is the time the element is operating during one cycle divided by the total cycle time. The fraction in this example would be 1.5 minutes ÷ 2.67 minutes = 0.56. In either case, whether you estimate or measure the time, you should wait about 10 minutes after the preheat cycle has ended before measuring the time. The oven requires that long to become completely heated so that the element will turn on and off at regular intervals. During these first 10 minutes, the oven uses more energy than during any later 10-minute period—but to keep our calculations as simple as possible, let's ignore this little extra energy.

When estimating the fraction of time the bake element is on, remember that you want the fraction of the total on-time; so if the element is on for about the same length of time it is off, it is operating half, or .50 (50 percent), of the time; if it is on for only about half as long as it is off, it is operating about a third, or .33 (33 percent), of the time. The heating elements do not glow red instantly when they turn on, and they don't lose their red color instantly when they turn off—so to find the on-and-off time, you must listen carefully for the relays clicking in the oven controls. Or watch your electric meter and time when the disc suddenly starts and stops.

To calculate the amount of energy needed to preheat your oven to 350° F. and maintain that temperature, fill in the appropriate blanks in table C-64 (Appendix C). Find the preheat energy by multiplying the time each element (bake or broil) is operating by the wattage of each. To convert the time in minutes to hours, however, we must divide by 60; and to convert watt hours to kilowatt hours, we must divide by 1,000. Suppose your bake element has a wattage of 2,000 and is operating for 11.42 minutes, and your broil element has a wattage of 3,000 and is operating for 2.75 minutes. The preheat energy for the bake element would be 2,000 watts × 11.42 minutes ÷ 60,000 = 0.38 kwh; and preheat energy for the broil element would be 3,000 watts × 2.75 minutes ÷ 60,000 = 0.14 kwh. The total preheat energy is the sum of these two energies, or 0.52 kwh.

You could repeat these procedures for finding the amount of energy needed for preheating and maintaining each different oven temperature that you use. Since that's a lot of work, however, you may want to take the easy way shown in table C-68 (Appendix C). Just multiply the amounts of preheat energy and power needed to maintain 350° F. by the appropriate factors in that table to find the amounts required for preheating and maintaining other temperatures.

To calculate the amount of energy needed to prepare various foods in your oven, just multiply the time (in hours) required for baking or cooking each food by the amount of energy needed to maintain the desired cooking temperature. Then add the preheat energy to find the total energy. Calculate the amounts of energy needed to prepare several of your most frequent dishes, and record these amounts in table C-69 (Appendix C). Remember, all times must be in hours!

Gas Oven One burner generally functions for both baking and broiling, operating constantly during the preheat cycle—but after the oven temperature reaches the desired value, the burner shuts off. The burner turns on and off periodically during the baking time to maintain the desired oven temperature.

Record the amount of gas consumption listed for your oven (either in the instructions or on the oven itself) in the proper column of table C-65 (Appendix C), and proceed from there. If this information is not available, determine the amount of gas consumption during the preheat cycle by using the test dial method (see chapter 2). Record the time for one complete revolution of the half-cubic-foot dial in table C-65 (Appendix C), then calculate the amount of gas consumption in cubic feet per hour.

Now you need to time the actual operation of the burner. Measure these times during actual baking conditions. Prepare some food that bakes at 350° F., and record the time in table C-66 (Appendix C) that the burner is operating during the preheat cycle. Then estimate the fraction of the time that the burner is on while the food is baking—or you can measure when the burner turns on and off, then calculate the fraction of on-time. For example, suppose the burner is on for 1 minute 30 seconds (1.5 minutes) and off for 1 minute 10 seconds (1.17 minutes); the cycle time is thus 2.67 minutes (1.5 + 1.17). The fraction of burner on-time is its on-time during one cycle divided by the total cycle time. In this example, the fraction would be 1.5 minutes ÷ 2.67 minutes = 0.56. In either case, whether you estimate or measure the time, you should wait about 10 minutes after the preheat cycle has ended before measuring the time. The oven needs that long to become completely heated so that the burner turns on and off at regular intervals. During these first 10 minutes, the oven uses more energy than during any later 10-minute period—but to keep our calculations as simple as possible, let's ignore this little extra energy.

When estimating, remember that you want the fraction of the total burner on-time; so if the burner is on for about the same length of time it is off, it is operating half, or .50 (50 percent), of the time; if it is on for only about half as long as it is off, it is operating about a third, or .33 (33 percent), of the time. Watch your gas meter test dials if you can't hear when the burner turns on and off, and record when the 1/2-cubic-foot dial suddenly starts and stops.

To calculate the amount of energy needed to preheat your oven to 350° F. and maintain that temperature for one hour, fill in the appropriate blanks in table C-67 (Appendix C) and do the calculations.

You could repeat these procedures for finding the amount of energy needed for preheating and maintaining each different oven temperature that you use. Since that's a lot of work, however, you may want to take the easy way shown in table C-68 (Appendix C). Just multiply the amounts of preheat energy and power needed to maintain 350° F. by the appropriate factors in that table to find the amounts required for preheating and maintaining other temperatures.

To calculate the amount of energy needed to prepare various foods in your oven, just multiply the time (in hours) required for baking or cooking each food by the amount of energy needed to maintain the desired cooking temperature. Then add the preheat energy to find the total energy.

Calculate the amounts of energy needed to prepare several of your favorite dishes, and record these amounts in table C-69 (Appendix C). Remember, all times must be in hours!

Microwave Oven The operation of a microwave oven is quite different from that of a conventional electric oven. A conventional oven uses electricity to heat an element, which heats the air in the oven, which then heats the food—a three-step process. And the heated air in the oven also heats the air in the kitchen by conduction through the oven walls and by convection of air through the air vent. In a microwave oven, the microwave generator converts the electricity to microwaves which heat the food—a two-step process. The one missing step is the heating of air in the oven—so a microwave oven loses none of its energy through the oven walls or air vents.

Microwaves have a rather unique set of characteristics that make them ideal for cooking. The primary characteristic, of course, is the fact that food substances and water absorb them and become hot—necessary condition for cooking. But microwaves also reflect off a metal surface—so if the inside surface of the oven is lined with metal, all the microwaves remain in the oven, bouncing off the walls until they are absorbed by the food. Microwaves also pass through glass and most ceramics, so the cooking dishes don't even get hot in the cooking process. (If the food cooks for a long time, however, heat from the food may heat the dish or pan.) The one problem with microwave ovens is that you can't use metal cooking pans, only glass or ceramic.

Although a microwave oven is 100 percent efficient in converting the microwaves to heat in the food, it is not 100 percent efficient in converting electrical energy to microwaves. Most microwave ovens have 450 to 700 watts of cooking power, but require 1,000 to 1,700 watts of input power. The actual efficiencies range from 40 to 50 percent.

It is difficult to define an efficiency for a conventional oven because of preheat times and the fact that a conventional oven can cook several items at the same time with little additional energy. We can, however, compare the energies required to cook various foods in conventional and in microwave ovens. Table 37 lists a few foods. Note that for most foods, the microwave oven uses between one-quarter and one-half of the energy required for a conventional oven. One reason for this large difference is that a microwave oven needs no preheat time—turn it on and it starts to cook. Also, because all the microwave energy is directed at the food, the food takes less time to cook. The meat loaf in table 37 requires an hour and a half in a conventional oven but only about 15 minutes in a microwave oven. A 12-lb. turkey requires only one hour and 20 minutes in a microwave oven.

A microwave oven does not save as much when compared to a cook top. Because a large fraction of heat from the cook top heating element ends up in the food, it is nearly as efficient as a

TABLE 37
Energy Required to Prepare Various Foods in Microwave and in a Conventional Oven

Food	*Energy Required For Conventional Oven*[a] *(kwh)*	*Energy Required For Microwave Oven*[b] *(kwh)*
Baked chicken (1 ½ lb.)	1.5	0.5
Pork chops (4 stuffed)	1.8	0.4
Meat loaf (1 ½ lb.)	2.0	0.5
Fillet of sole (1 lb.)	0.9	0.2
Swiss steak (3 lb.)	3.2	1.2
Lasagne (10-inch pan)	1.5	1.3
Turkey (12 lb. not stuffed)	4.5	1.9
Frozen pizza (14-inch)	0.8	0.2
Baked potato (4 potatoes)	1.8	0.5
Sweet potatoes (4 servings)	2.2	0.3
Green vegetables (4 servings)	0.3[c]	0.2
Squash (acorn or butternut)	1.6	0.3
Stewed tomatoes	0.7[c]	0.2
Carrots (4 cups)	0.3[c]	0.2
Quick breads	1.5	0.2
Corn bread	1.0	0.2
Coffeecake	1.2	0.3
1-layer cake	1.0	0.2
2-layer cake	1.1	0.3
Apple pie (9-inch)	1.8	0.4
Pumpkin pie (9-inch)	1.5	0.5
Pineapple upside-down cake	1.2	0.3

[a] Energy values for a conventional oven are based on a preheat energy of 0.5 kwh and 1.0 kw power to maintain 350° F. Times and temperatures are taken from some popular cookbooks and box directions. If two or more items are prepared in the oven at one time, the total amount of energy required is just slightly more than the amount required for the item that takes longest to cook.

[b] Energy values for a microwave oven are for a 675-cooking-wattage unit that uses 1,450 watts of input power. Cooking times are taken from a microwave cookbook and the author's own experience. If two or more items are prepared in the oven at one time, the total amount of energy required is the sum of the amounts required for each.

[c] These values are for preparations on a cook top rather than in an oven.

microwave oven. The three items usually prepared on a cook top that are listed in table 37 show the small comparative difference in energy requirements for cooking.

But how does food cooked in a microwave oven compare to that prepared in a conventional oven? Baked goods and meats tend to come out more moist in a microwave because they aren't exposed to high heats for a long time—on the other hand, they don't brown as much. In fact, any food cooked for less than half an hour probably won't brown. (But a turkey, which cooks for more than an hour, does brown nicely, not quite the deep brown you get in a conventional oven, but it is brown.) So some foods don't look the same and some don't quite taste the same—and whether they turn out better or worse than if prepared in a conventional oven *is,* finally, a matter of taste. You'll also find that not only do foods come out differently, but that you'll need to revise your cooking schedule. Because of the short cooking times, your dinner will be prepared before you're ready to sit down to eat.

Calculations for the amounts of energy required to cook food in a microwave oven are much easier to figure than calculations for a conventional oven, especially if you have only one cooking power. Find the input power, usually listed on the back of the unit, or measure it with the rotating disk method (see chapter 2). Record this value in table C-70 (Appendix C). Then multiply the total cooking time (in hours) by the input power to find the amount of energy used to cook each kind of food. Use table C-69 (Appendix C) for your calculations, but ignore the blanks for preheat energy.

You can figure one of several ways if your microwave oven has variable cooking speeds. If you use maximum power most of the time and the lower settings infrequently, just assume that the oven uses maximum power at the lower settings also; the error in the total amount of energy used per month will be small. Or actually measure the power at the different settings. To attain these variable cooking speeds, the microwave generator turns on and off at short intervals—two seconds or less. You will see that your electric meter disc turns a little, then stops, turns a little more, stops, and so on, as the power turns on and off; the disc may start and stop ten to twenty times during one revolution. These stop-start intervals are short enough to enable you to time one revolution of the disc. Use that value to find the effective power of the oven at the lower cooking speeds. Or just measure the effective power at the lowest setting and make an educated guess at the effective power of the other settings.

Once you have values for the effective power at the other settings, you can find the amount of energy required to cook food at these settings. Calculate this amount in table C-69 (Appendix C)—except use effective power instead of maximum power figures. If certain foods require two or more different power settings, calculate the amount of energy used at each setting, then add them to find the total energy amount. For example, suppose you cook a food for 5 minutes (0.08 hour) at a low setting that uses 0.7 kw of power; and 8 minutes (0.13 hour) at a high setting that uses 1.5 kw of power. The energy at the lower setting is 0.7 kw × 0.08 hour = .06 kwh; and the energy at the high setting is 1.5 kw × 0.13 hour = 0.20 kwh. The total amount of energy needed to cook the food would be 0.06 + 0.20 = 0.26 kwh.

Cook Tops

Most cook tops have four electric heating elements or gas burners. An electric cook top has one or two large heating elements with a maximum output of 2,100 to 2,500 watts and two or three smaller ones with a maximum output of 1,500 to 1,800 watts. A gas cook top has one or two high-heat burners that produce a maximum of 10,000 to 12,000 Btu per hour (and use 10 to 12 cubic feet of gas per hour) and two or three medium-heat burners that produce a maximum heat of 8,000 to 9,000 Btu per hour (8 to 9 cubic feet per hour). All cook tops have a temperature control for each burner or heating element—either a continuous dial for gas and some electrics, or push buttons for other electrics.

A continuous dial on a gas burner controls gas flow to the burner and thus controls the heat. The control of electric heating elements is not as simple. The electric current flowing through a heating element can't be varied continuously; it is either on or off, or by means of special electrical circuitry can be set at selected values between no current and maximum current. Electric cook tops with push-button controls have this special circuitry, and the current through a heating element can be set at four or five specific values. So the heat produced by an electric heating element with push-button control can only have four or five preset values; nothing between these values is possible.

Electric heating elements with continuous control use a bimetallic heating device that turns the electric current on for a few seconds, then off for a few seconds. For low heat settings, the electricity is on for a short period, then off for a longer period; the electricity for high heat settings is on for a long period, then off for a short period. As you turn the control from a low to high heat setting, the on-period becomes longer and the off-period shorter.

You can probably hear the relays in a continuous-control electric cook top turn the electricity on and off.

To find the amount of energy needed to cook a given food on your cook top, you must know the amount consumed by the heating element or burner and the time needed to cook the food. The maximum output energy of each heating element or burner should be listed in the cook top instructions. List these numbers in the appropriate blanks in table C-71 (Appendix C). If the maximum values aren't listed or you just can't find them, use the rotating disc or test dial method (see chapter 2) to find the wattage or cubic feet per hour consumed by your cook top.

Now what about the settings below maximum? Most of your foods are cooked at a medium to low setting, and the best way to find the amount of energy used at these settings is to measure the actual energy consumed at all of the lower settings—but who wants to go through all that effort? The easiest way is just to estimate the amount of energy. But a heating element or burner still produces about 15 percent of its maximum output on the lowest setting (simmer), so the heat output of the element or burner is not proportional to the control-dial setting. For example, if the control dial is set halfway between the minimum and maximum setting, the output of the burner is about 60 percent of the maximum output, not 50 percent; and if the dial is set at about one-fourth between minimum and maximum settings, the element or burner output is about 35 percent of the maximum output.

Estimate the fraction of the maximum output for your burners or heating elements at a lower setting by referring to table 38. For push-button electric cook tops, simply read the fraction of the maximum output for each button, and record those values in table C-72 (Appendix C). For example, if you have a five-button control (six buttons including the "off-button") and depress the third control button, the heating element would produce 0.60 (60 percent) of its maximum heat. For continuous-control cook tops, determine which number on the dial represents about one-eighth of the way from low to high setting, which number represents about one-fourth, three-eighths, and so on, of the way. The fractions of maximum heat for these numbers are given in table 38. For example, if the numbers on a dial range from 1 for the lowest setting to 15 at the highest setting as shown in figure 24, the number 8 is halfway from the lowest to highest setting; the num-

TABLE 38
Fraction of Maximum Heat for Stove-Top Controls at Different Settings

Push-button Control:

Four-button control

Button Number	*Fraction of Maximum Heat*
0 (off)	0.00
1 (low)	0.15
2	0.45
3	0.70
4 (high)	1.0

Five-button control

Button Number	*Fraction of Maximum Heat*
0 (off)	0.00
1 (low)	0.15
2	0.35
3	0.60
4	0.80
5	1.0

Continuous Dial Control:

Fraction of Dial Setting From Minimum to Maximum	*Fraction of Maximum Heat*
0 (minimum)	0.15
⅛	0.25
¼	0.35
⅜	0.45
½	0.60
⅝	0.70
¾	0.80
⅞	0.90
1 (maximum)	1.0

ber 4 (or 5) is about one-fourth of the way; and so on. The burner or heating element with the dial set at 8 would produce about 0.60 (60 percent) of the maximum heat; at the number 4, it would produce about 0.35 (35 percent) of the maximum heat; and so on. Record the dial settings you most frequently use in table C-72 (Appendix C), then record the fractions of maximum output.

Calculate the effective wattage or cubic feet per hour in table C-72 for each setting by multiplying the fraction by the maximum wattage or cubic feet per hour. For example, if the maximum output of a burner is 12 cfh and the dial of our example is set at 10 (about five-eighths of the way from LOW to HIGH), we find the fraction from table 38 to be 0.70; so the burner output is $0.70 \times 12 = 8.4$ cfh. Don't worry about the accuracy of this method, because you probably don't dial the same setting each time you cook anyway.

If you decide to measure the actual energy outputs for each setting, use the methods shown in tables C-63 and C-64 (Appen-

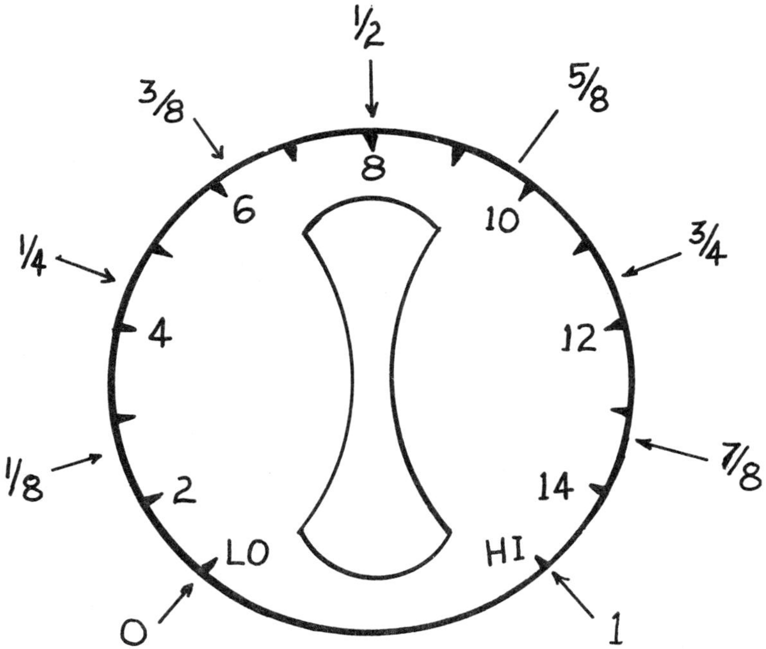

Fig. 24. Cook top dial

dix C) for electric cook tops, or use the test dial method (see chapter 2) for gas cook tops.

Many foods require two different settings on a cook top: a high-heat setting until the food boils, then a lower setting until finished. Thus you need to calculate the amount of energy used at both settings and add these two figures together to find the total energy. Fill in the blanks in table C-73 (Appendix C) for several of your most frequently prepared foods. All cooking times should be in hours (divide the number of minutes by 60). A vegetable that requires 7 minutes (0.12 hour) to come to a boil at the maximum heat of 9 cubic feet per hour, uses $9 \times 0.12 = 1.1$ cubic feet of gas. If the vegetable then requires 10 minutes (0.17 hour) to cook at a setting with 0.25 of the maximum heat output, the energy amount at this setting would be $9 \times 0.25 \times 0.17 = 0.4$ cubic feet. The total amount of energy required to cook the vegetable would be $1.1 + 0.4 = 1.5$ cubic feet.

Other Cooking Appliances

Except for gas grills that are part of cook tops and outdoor gas units, all other kitchen appliances are electric. For appliances that have thermostats, such as waffle irons, toaster ovens, and grills, you must time when the electricity is actually on, just as you did for the oven—and for some appliances you need to know the amount of preheat energy. Fill in your most frequently used temperature settings in table C-74 (Appendix C) for each appliance. For example, suppose your 1,500-watt (1.5 kw) electric grill requires 20 minutes (0.33 hour) to heat up to 350° F. and is on for 75 percent of the time after preheating to maintain that temperature. The preheat energy is $1.5 \times 0.33 = 0.5$ kwh; and the power required to maintain 350° F. is $1.5 \times 0.75 = 1.13$ kw.

Now calculate the energy required to prepare various foods on the different appliances. For appliances with thermostats, you must add the amount of preheat energy to the amount needed to cook the food. If you use the 1.5 kw grill of our example to make pancakes for 30 minutes, the energy required for these 30 minutes (0.5 hour) will be the amount needed to maintain 350° F. multiplied by the time, or 1.13 kw $\times$ 0.5 hour $=$ 0.6 kwh. The total amount of energy required to make the pancakes would be $0.5 + 0.6 = 1.1$ kwh.

Do the first calculation in table C-75 (Appendix C), then use that method to calculate the energy used to prepare your other most frequent foods. For foods prepared on appliances without a thermostat, the amount of energy is just the wattage of the appliance multiplied by the time the appliance is on.

In addition to cooking, you may also use some energy for mixers, blenders, or other small appliances. Include these calculations in table C-75 also.

Adding It All Up

Now that you have calculated the energy used to prepare the various foods you eat, the task of finding the total energy amount used in the kitchen becomes one of bookkeeping. The major problem is that most families have no weekly or monthly pattern of food preparation. Probably the best way to arrive at some reasonable estimate of a monthly energy use is to prepare a sample menu for one month—one that represents an average number of each of the different foods you prepare. For example, is spaghetti served once or twice a month? Are pork chops served once a month, more often, never? Don't forget to include the cakes, pies, cookies, and other goodies you prepare. Or just keep a list of the foods you prepare for one month. From this menu, list the foods and the number of times each is served during a month in table C-76 (Appendix C). Be sure to take into account the number of times you may eat out and the days that you may entertain guests and thus require more (or less?) energy to prepare meals.

Conservation Measures

One way to conserve energy in the kitchen is to stop eating, but most people have a habit of getting hungry after a few hours. Another alternative is to stop cooking, but constantly cold meals tend to become less appetizing. So you will probably have to cook unless you can afford to eat at restaurants all the time. You can prepare the same meals but use less energy, however, if you just think about how and where the energy is used and plan your cooking and baking with that in mind. The following suggestions should help you save some energy. Except in a few cases, calculating the energy saved is difficult because you can't obtain the exact information needed—so most of these suggestions won't include a method of calculating the energy savings.

During the heating season your actual savings will be less than calculated here, especially if you have electric heat. Refer to chapter 9 for details.

Use a Toaster Oven for Small Amounts of Food Most toaster ovens use about half the energy of a large oven to maintain a desired cooking temperature. And a toaster oven requires little preheat energy, so it can save from 50 to 80 percent of the energy you would use with a large oven. Cost of the energy saved with a toaster oven over an electric oven could pay for the toaster oven in ten years. If you have a gas oven and the price of your gas is relatively low, however, a toaster oven may not save you any money at all.

Turn Off Your Oven Before the Food Finishes Cooking An oven retains heat quite well, so turn it off 5 minutes before the food is finished. The heating elements on electric cook tops also retain heat for a minute or two, so turn them off just before the food is finished.

Don't Preheat Your Oven Except for foods that cook in a few minutes, don't bother to completely preheat the oven, but do allow a few extra minutes of cooking time. If your electric oven uses the broil element in its preheat cycle, be sure that element is off before you place any food in the oven.

Put as Much in Your Oven as Possible If you're cooking several items, place as many of them as possible in the oven at one time. Most of the heat goes toward heating the oven, so the extra food requires only a little extra heat. If the different items require slightly different temperatures, set the oven on the average temperature and cook the items that require lower heat for a slightly shorter time than required and the items that require higher heat a little longer than required. Or, if your oven space is limited, place the items in the oven one after the other to save the preheat energy for the second item. If part of a meal requires the oven and another part requires the cook top, can you use the oven to prepare the entire meal?

Defrost Food Before Placing It in the Oven Cooking time is less if the oven doesn't have to thaw the frozen food as well as cook it.

Don't Peek You want the heat to stay in the oven and cook the food, not warm your face. Each time you open the oven door, 20 percent of the inside heat escapes.

Use Glass or Ceramic Pans For baking, these are more energy-efficient than metal pans—you can lower cooking temperature by 25° F. and save 5 to 10 percent of the energy used at the higher temperature.

Cook in Bulk Prepare enough food for two meals at a time, then reheat the second portion another day. The amount of energy needed to reheat a food is less than that needed to cook it originally.

Broil Meats Broiling requires less energy than cooking meats in the oven.

Use Hot Water from the Tap When boiling or heating large quantities of water, start with hot water from the faucet. Your hot-water heater is more efficient than the cook top. Savings are greatest if you have an electric cook top and an oil or gas water heater. For small amounts of water, however, you must consider the heat losses along the pipe from the hot-water heater—if such losses are large, you would do better to heat cold water on the cook top.

Use a Minimum of Water More energy is required to heat up unneeded water, so use as little as possible when cooking. Vegetables need only enough water to cover the bottom of the pan. Heat only the amount of water you need for instant coffee or tea, not a whole kettle.

Use a Pressure Cooker Pressure cookers use less energy than regular cook top preparation.

Use Flat-bottom Pans Be sure your pans have flat bottoms, especially when you're using an electric cook top. A warped pan means a bad contact with the heating element—heat is not efficiently conducted to the pan. The pan should cover the entire heating element, as heat from any uncovered part of the element goes to the room, not to the pan. For gas burners, don't let the flame go up the sides of the pan—most of that heat also goes to the room.

Don't Waste Half Your Toaster Energy A toaster uses the same amount of energy toasting one slice of bread as it does in toasting two—so if you need two pieces of toast, toast them together.

Don't "Over-boil" Water Boiling water has the same temperature whether it's boiling rapidly or slowly. The only difference is that the rapid boiling water produces more steam from the extra energy used. When heating water to a boil, keep the lid on the pot; the water will come to a boil faster. Once the water boils, reduce heat to a setting just sufficient to keep it slowly boiling.

Planning to Purchase a New Oven? Consider buying a microwave oven. Refer back to the section in this chapter describing microwave ovens for advantages and disadvantages. Also talk to people you know who own a microwave to find out their experience. Remember that one advantage is the faster cooking times; so if you have a busy household and only limited time to prepare meals, a microwave oven could be a convenience that actually uses less energy, for once, than the "old" way of doing things. A second advantage, for those who don't like to wash dishes, is that food doesn't bake onto the pan because the pan doesn't get hot during the cooking process.

Energy Savings You can calculate the expected energy and dollar savings in table C-77 (Appendix C) for the foods you can prepare in a microwave oven. This calculation can give you a rough idea of how much you would save in preparing the same foods that you now prepare. With a microwave oven, however, you may find that you can prepare different foods more easily, so your menu may change. Don't expect the savings to pay for the microwave oven over a ten-year or fifteen-year period unless you presently do a lot of cooking in a large, inefficient oven. Microwave ovens are expensive, but the added convenience plus savings may make it worthwhile for you.

If you aren't ready for a microwave oven, consider a self-cleaning oven. The oven heats to a very high temperature during the self-cleaning cycle, and food on the oven walls turns to a char that you can easily wipe away. In order to make the oven safe, the manufacturers have installed better insulation in the oven walls so the outside of the oven doesn't get too hot. This better insulation also retains heat in the oven while cooking, so the oven loses less heat to the room and uses less energy than a conventional oven to cook your food. Each cleaning of a self-cleaning oven, however, requires 3 to 6 kwh of electricity or 30 to 60 cubic feet of gas; so if you use the self-cleaning feature on the recommended schedule, the total consumption of energy will be more than for a conventional oven because of the extra heat needed for the cleaning cycle. If you use the self-cleaning feature sparingly or not at all (cleaning the unit like a regular oven), the self-cleaning oven can save energy.

Do not confuse a self-cleaning oven with a continuous-clean oven. A continuous-clean oven has a coating on the oven walls that food doesn't readily stick to. It has no more insulation than a conventional oven and won't save any energy.

If you still prefer a conventional oven, check the relative efficiencies of the various models. In a past survey, *Consumer Reports* rated several ranges and found an annual difference in energy consumption of about 200 kwh between the most and least efficient electric models. At 5 cents per kwh, this means an annual savings of nearly $10, or about $150 over the fifteen-year life of the range. Gas ranges differed by as much as 14 Ccf per year for an annual difference of over $4, or about $60 for fifteen years. No present listing of relative efficiencies for all oven models is available, so this information may be hard to obtain. Your best option is to check consumer magazines that periodically rate various oven models for relative efficiencies. Or be guided by the thickness of the insulation in the oven walls—the thicker the insulation, the less heat loss and energy used. If the salesperson doesn't know the insulation thickness, don't buy the oven until you find out.

When buying a gas oven or cook top, choose one with an electronic or mechanical ignition rather than a pilot light. The amount of electricity required for the ignition system is negligible, so the amount of gas that would be used to keep the pilot light burning is an energy saving. Use table C-78 (Appendix C) to calculate the expected savings.

Dishwasher

Dishwashers come in two basic types. One type has an electrical pump that sprays hot water to wash the dishes, and the other has two hands and a dishcloth to wash the dishes. The best type for your home depends on the cost of your electricity and hot water and the effort required to get the hands into the dishwater. We will be primarily concerned in this section with the first type. We will also discuss the costs of washing dishes by hand (but not how to overcome the problem of getting someone to do them).

A dishwasher has a pump, an impellor, or both, which generate a fast hot-water spray to wash the food off the dishes. A heating element heats the water to a temperature higher than that of the entering hot water, so a dishwasher uses electrical energy for the motor and for the heating coil. The total amount of energy used by the dishwasher consists of the electrical energy required for the pump and heating coil and the energy used to heat the incoming hot water.

How Much Energy Does Your Dishwasher Use?

Most dishwashers have several cycles: an initial rinse cycle, a wash cycle, probably another rinse and wash cycle, one or two more rinse cycles, then a dry cycle. Many newer models have a rinse/hold cycle: if you wash your dishes only once a day, you can use the rinse/hold cycle to rinse the dishes after each meal. To calculate the amount of energy used by your dishwasher, you must find the wattage of the heater and the pump motor, the time they operate during the different cycles, and the amount of hot water used for each cycle. Generally the heater and pump motor always operate together, although in some models the motor is on and the heater off during the water fill—or the motor is off and the heater on. But to keep the calculation simple, let's assume that motor and heater are always on together and the wattage of the dishwasher equals the wattage of motor and heater combined. Because wattage of the heater is much higher than that of the motor, we will consider the heater operating time as the operating time for the dishwasher and ignore the times that the motor operates by itself. This assumption will mean that our calculated energy use will not be exact, but the calculations will be simpler than trying to deal with the wattage and times of heater and motor separately.

Both wattage and times should be listed in your instruction manual. If you have your instruction manual, refer to the following section; but if you don't have the manual or if it doesn't list the heater wattage, you must do a little more work, so refer to the section following the next. The instruction manuals for some new models have the total energy use listed for each different setting. For these dishwashers you can just record the kilowatt hour and hot water amounts in tables C-80 and C-81 (Appendix C).

Instruction Manual Available Find the total amount of time from the instructions that the heating element is on for the cycles that you use, including any rinse/hold cycles. Some dishwashers have a hold phase during some wash, rinse, or both cycles. For these cycles, the timer stops and the heater remains on to heat the water to a prescribed temperature. Then the timer starts again. If your dishwasher has a hold phase in a cycle, you must measure the actual time of that cycle to find the on-time of the heater.

Find the wattage of the heating element, and record that value in the appropriate space in table C-79 (Appendix C). The motor wattage probably won't be listed, although the horsepower (hp) rating may—but how do you convert horsepower to watts? Basically you can't; the wattage depends on the load of the motor, so unless you know the engineering specifications of the appliance, you can't easily make the conversion. You can, however, use the approximation that a 1/5-hp motor uses 120 watts, a 1/4-hp motor uses 150 watts, a 1/3-hp motor uses 200 watts, and a 1/2-hp motor uses 300 watts. Or you can actually measure the wattage.

If you need or want to measure the wattage, use the rotating disc method (see chapter 2). Because you actually want the total wattage of motor and heater together, don't attempt to measure the wattage of just the motor; measure the wattage when heater and motor are on together (during a wash or rinse cycle after the fill is complete). This is the wattage of the dishwasher. Record the wattage values in table C-79 (Appendix C). The amount of energy needed to run the drain pump and water valve is small and you can ignore it.

No Instructions Available You must find two values: the on-time of the heater and the wattage of heater and motor together. To find the wattage of motor and heater, use the rotating disc method. Measure the wattage after the water fill has finished, and record the data in table C-79 (Appendix C).

The on-time of the heater should start at either the beginning or end of the water fill. The pump motor also turns on at either of these times, so the on-time of the motor is not necessarily the on-time of the heater. Time when the heater turns on by watching the disc on your electric meter just before and during the water fill. If the disc rotation increases sharply when the fill starts, the heater turns on at the beginning of the water fill; if the rotation increases at the end of the fill, the heater turns on then. The motor also turns on at the beginning or end of the fill, somewhat increasing the disc rotation, so be sure that any increase in rotation is due to the heater turning on. Or just measure the wattage during and after the water fill and compare the two values.

For each cycle, time from when the heater comes on, either at the beginning or end of the water fill, to when the spray ends (when the pump starts to empty the water). Add the times for the different cycles together, then record the value on table C-79 (Appendix C). Include any rinse/hold cycles that you may use during the day. The amount of energy needed to operate the drain pump and water valves is small and can be ignored.

Now that you know the wattage and times, use table C-80 (Appendix C) to calculate the total amount of electrical energy used by your dishwasher. For example, if the heater is on for a total of 35 minutes (0.58 hour) for all cycles and the power of heater and motor together is 1,300 watts, the electrical energy required for one use of the dishwasher is 1,300 watts × 0.58 hour ÷ 1,000 wh per kwh = 0.75 kwh.

To find the amount of hot water used by your dishwasher, find the number of quarts per fill as listed in the instruction manual. These values are for pressures in a municipal water system (40 to 60 lb. per square inch). If your water comes from a well and the maximum pressure is lower than 50 to 60 lb. per square inch, or if the water pressure from your municipal system is low, the amount of water may be less than what is specified in the instructions. If you have the instructions and have normal water pressure, record the listed water fill in table C-81 (Appendix C), and continue from there. If you have a lower water pressure or don't have the instruction manual, you will have to determine the amount of hot water actually used per fill. Most new models fill to a specified volume rather than for a specified time. For these models, the amount of water per fill is independent of the water pressure.

Open the door of the dishwasher during a wash or rinse cycle after the water fill is completed and note the level of water in the bottom of the dishwasher. Then later, when the dishwasher is empty, repeatedly fill a quart container with water (cold water will suffice—don't waste energy) and empty it into the dishwasher until the water level is the same as it was during the wash or rinse cycle. Record the number of quarts required to fill the dishwasher in table C-81 (Appendix C). Then set the controls to drain the water from the dishwasher (or you may be washing the floor the next time you use the dishwasher).

Now that you know the amounts of hot water and electrical energy required per use of the dishwasher, use table C-82 (Appendix C) to find the monthly energy consumption and costs. If the dishwasher of our example uses 15 gallons of hot water along with 0.75 kwh per use, it would require 450 gallons of hot water and 23 kwh for thirty uses per month. Cost of the electricity would be $1.40 per month (at 6 cents per kwh), and cost of the hot water would be about $1.60 per month if heated by a gas or oil hot-water heater—about $5.25 per month if heated by an electric hot-water heater.

Dishwasher or Hands—Which Uses Less Energy in the Soapsuds Derby?

What about washing dishes by hand rather than by dishwasher—does the dishwasher use more or less energy? The answer is yes, no, or maybe, depending on how you wash the dishes by hand. If you keep the rinse water flowing constantly, hand washing uses about the same amount of energy as a dishwasher; but if you turn the rinse water on and off to rinse each dish or fill another sink or dishpan with rinse water and dip the dishes into it, hand washing uses less energy than a dishwasher.

But these results apply only to one hand washing contrasted to one dishwasher operation for a full load of dishes. Families who own a dishwasher generally load it with dishes after each meal but wash all the dishes only once a day. Families without a dishwasher, however, generally wash the dishes after every meal, so probably use more energy than if they washed all of the day's dishes just once.

The task of finding the amount of energy used to wash dishes by hand is not simple. You probably have different water temperatures for the wash and rinse water, you change the flow and temperature of the rinse water while washing, and you are not generally consistent about the amount of water used. Thus we will have to make some assumptions in order to simplify the calculations.

Let's assume that either the rinse water is the same temperature as the wash water (let's call this the "warm water"), or that the rinse water is only hot water (from the hot-water faucet only). The warm water is part hot and part cold water, but we need to know how much of this warm water is hot water from your hot-water heater. Let's call the part that is hot water the "hot-water fraction"; so the task of finding the total amount of hot water used to wash dishes by hand means finding the hot-water fraction and the total amount of warm water used.

To find the hot-water fraction, measure the temperatures of the hot water, the cold water, and the warm water that you use for washing. The hot-water fraction of the warm water is the temperature difference between the warm and cold water divided by the temperature difference between the hot and cold water. Record these temperatures in table C-83 (Appendix C) and do the calculations. Remember to measure all temperatures with the thermometer in the *flowing* water and to let the hot water flow long enough to flush out the cooled water in the pipe. As an example, suppose your hot water has a temperature of 140°, your cold-water temperature is 60°, and the wash-water temperature is 120° F. The temperature difference between warm and cold water is 60° F. (120° − 60° = 60° F.), and the temperature difference between hot and cold water is 80° F.; so the hot-water fraction is 60° ÷ 80° = 0.75.

Now you must find the total water volume used for the wash and the rinse. To determine the amount of wash water, repeatedly fill a quart container and pour the water into the sink or dishpan a quart at a time. Count the number of quarts needed to fill the sink or dishpan.

Measuring the rinse water may not be as easy. If you use a pan or sink for rinse water, fill the pan or sink a quart at a time, counting them as you did with the rinse water. But if you rinse with flowing water from the faucet, you have more work to do. For a double sink, place a container under the rinse water in the second sink and measure the amount of water in the container when the dishes are finished. If you have only one sink and the rinse water flows into the wash water, measure the initial level of the wash water, in inches, from the bottom of the sink. Then, after washing the dishes, use a quart, two-cup, or other measure to dip out the water until the level reaches the original wash-water level. The amount dipped out is the amount of rinse water. Record the wash-water and rinse-water volumes in table C-84 (Appendix C), then calculate the amount of hot water used to wash your dishes by hand.

Conservation Measures

Dishwashers are designed to wash all kinds of dried and burned-on food off the dishes. Most families don't have exceptionally dirty dishes to wash every day, however, so a dishwasher is overdesigned for normal needs and may use more energy than is really needed to wash the dishes satisfactorily. Thus you don't need the full capacity of your dishwasher, and you can save energy by using only the capacity needed to wash your dishes.

The following suggestions could help reduce the energy consumption of your dishwasher.

Use Only One Wash Cycle; Don't Use the Rinse/Hold Feature Don't use two wash cycles unless absolutely necessary for badly soiled dishes. Use just one wash with a prerinse cycle and one or two rinses after the wash. If the dishwasher doesn't clean off some dishes with just one wash, rinse those dishes before placing them into the dishwasher. Eliminate the rinse/hold cycle by hand-rinsing the dishes, using cold water except for greasy dishes.

Energy Savings The energy saved is the amount of electricity and hot water that would be used for the cycles you eliminate. Use table C-85 (Appendix C) to calculate your expected energy savings. If you can eliminate a wash and rinse cycle that use 2 gallons of hot water and 0.1 kwh each, your monthly savings would be 120 gallons of hot water and 6 kwh of electricity. The money saved would be 30 cents for the electricity and 40 cents (gas or oil hot-water heater) to $1.20 (electric hot-water heater) for the hot water. Your annual savings would range from nearly $8 to $17.

Eliminate the Dry Cycle Stop the dishwasher after the last rinse and open the door, as the dishes are hot enough to dry adequately. Some dishwashers aren't easy to reset on start if the dry cycle is not used; the instruction manual or your appliance dealer may show you how to bypass the dry cycle. Most new dishwashers have a "power-saver" setting that automatically eliminates the dry cycle, so this task is much easier.

Energy Savings The amount of energy saved is the amount of electrical energy needed to run the heater and motor for the duration of the dry cycle. If the dry cycle is 20 minutes long (0.33 hour) and the power is 1,300 watts, the amount saved would be 1,300 watts × 0.33 hour ÷ 1,000 wh per kwh = 0.43 kwh. The energy saved per month (30 uses) would be 13 kwh with the resulting saving of 50 cents per month or $6 per year. Use table C-86 (Appendix C) to calculate your savings.

Fill Up Your Dishwasher When using your dishwasher, be sure it is full but not overloaded. Washing a half load uses the same amount of energy as washing a full load.

Planning to Buy a New Dishwasher? Compare the different

models for energy use. Ask the salesperson for the instruction manual. The instructions or advertising brochures for some new models have the actual amounts of electrical energy and hot water used for the different cycles. If no information about the energy consumption is listed, use the methods outlined in the previous sections to calculate the amounts of electrical energy and hot water used by each. Because the amount of energy needed to run the motor is approximately the same for all models, find the heater wattage and the on-time of the heater, then compare the costs of heater operation and hot water for the different models. Remember that 0.1 kwh of electricity and 1 gallon of hot water saved per day amounts to 36.5 kwh and 365 gallons of hot water per year. These energy reductions would mean a saving of $3 per year, or about $45 over fifteen years, for oil or gas hot-water heaters; or a saving of about $5 per year, or nearly $75 over fifteen years, for electric hot-water heaters.

6.

Air Conditioners and Dehumidifiers

An air conditioner or dehumidifier can use a substantial amount of energy during the summer months. If you live in one of the southern states, your cost for air conditioning during the summer can be greater than your cost for heat during the winter. You use both of these appliances mainly for your comfort, so the amount of energy consumed depends on how comfortable you want to be. The energy consumption also depends on the efficiency of the air conditioner or dehumidifier. Of all appliances, air conditioners have the widest range of efficiencies. Therefore you must purchase an air conditioner or dehumidifier with care, and you need to be conscious of how you use them if you want to keep energy consumption at a minimum.

Air Conditioner

An air conditioner is just a heat pump that removes heat from inside the house and sends it outside. (See chapter 3 for the description of a heat pump.) A window unit is shown in figure 25.

In addition to cooling the room air, an air conditioner also removes moisture from the air. As the air passes over the cold coils, the water vapor in the air condenses on the coils, drips into

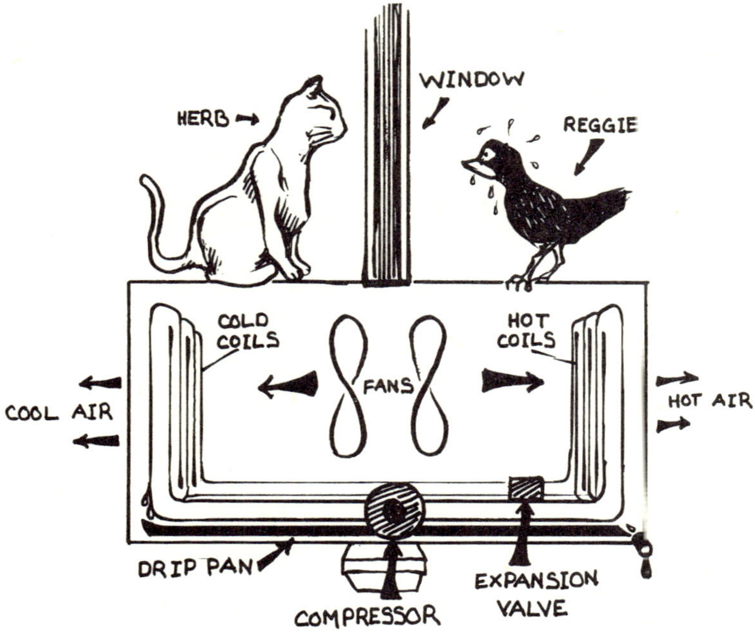

Fig. 25. Typical window unit air conditioner

a pan, then flows outside through a small opening. Each pound of water condensed (about one pint) releases 1,000 Btu of heat, which the air conditioner must remove; so in regions where the humidity is high, the amount of energy used just to condense water vapor can be a significant part of an air conditioner's load. A 10,000-Btu-per-hour air conditioner that removes 2 pints of water per hour produces only 8,000 Btu per hour of cooling—the other 2,000 Btu per hour are required to condense the water vapor.

The amount of energy required to cool a house depends on three factors: the rate of heat gain through the walls, doors, windows, and ceiling of the house; the amount of moisture in the air; and the efficiency of the air conditioner. The calculations for heat gains are the same as those for heat losses during the heating season—except, of course, the heat is flowing into, rather than out of, the house. Yet the calculations for heat gains are much more difficult. The greatest problem with calculating expected heat gain during the cooling season is the large variability of the outside temperature. During most of the heating season, the outside temperature is cold enough to require almost constant operation of the heating system. But during the cooling season, outside temperature can be cool enough for a significant part of the time to require no air conditioning. This "on-and-off" situation means that an average length of cooling season and an average outside temperature are hard to define; thus an average heat gain is also hard to calculate.

In addition, the effect of heating by direct solar radiation is small during the heating season; but during the cooling season, this is a major source of heat input to the house. Because the heat from solar radiation depends on such things as the color of outside walls and roof, the shading of nearby trees, the length of the eaves, and the amount of direct sunlight received every day, this heat input is nearly impossible to calculate without specific details for each house. So all we can do in the following section is to calculate the amount of energy you now use for air conditioning, suggest ways to reduce this energy consumption, and make a few rough estimates of how much energy you can save.

Relative Efficiencies and Costs

We defined an appliance's efficiency as the useful work of the appliance divided by the energy input:

$$\textbf{Efficiency} = \frac{\textbf{useful work}}{\textbf{energy input}}$$

The useful work of an air conditioner is the amount of heat it removes from the house (measured in Btu). The energy input is the amount of electrical energy used to remove that heat (measured in watt hours rather than kilowatt hours). Therefore the efficiency is:

$$\textbf{Efficiency} = \frac{\textbf{Btu removed}}{\textbf{Watt hours consumed}}$$

This value is known as the EER (Energy Efficiency Ratio) and is measured in units of Btu per watt hour (Btu/wh). The EER values of air conditioners range from about 5 to almost 12 Btu/wh. Most air conditioners have the EER value listed; but if not, you can easily calculate it by dividing the Btu per hour (written as BTUh) cooling capacity by the wattage[1] of the unit. For example, an air conditioner with a cooling capacity of 10,000 Btu/h and a power of 1,500 watts has an EER of 5.6 Btu/wh (10,000 Btu/h ÷ 1,500 watts).

If the EER value of your air conditioner is listed on the unit, enter that value in the appropriate blank in table C-87 (Appendix C); but if not listed, calculate the value in the table. If the wattage isn't listed, use the rotating disc method to find it (see chapter 2).

To compare the costs of operation for air conditioners with different EER values, let's consider the ten-year operation costs of a 10,000-Btu/h unit that operates for 800 hours per summer (the estimated number of cooling hours needed for Washington, D.C.—excluding the hot air produced in certain locales of the city). The costs of operating the various units are listed in table 39 for different electrical costs. As you can see, the cost of operating a low EER unit is much more than that for operating a high EER unit. Generally a high EER air conditioner costs more to purchase than a low EER unit, but you can see from table 39 that the lower operation cost of the high EER unit can be worth the extra initial purchase price.

Air conditioners with an EER of 11 to 12 Btu/wh are not very plentiful on the market, so you may have a difficult time finding one. You should be able to find a unit with an EER of 9 to 10 Btu/wh without too much effort, but you'll have to search among many units with EER values of 6 to 8 Btu/wh. From table 39 you can see that at 5 cents per kwh, a unit with an EER of 10 Btu/wh can save more than $250 during the unit's expected ten-year life when compared to a unit with an EER of 6 Btu/wh—a saving comparable to the cost of an air conditioner.

TABLE 39
Costs of Operating Various 10,000-Btu/h Air Conditioners for 800 Hours per Season

EER	*Watts*	*kwh/ year*	*kwh/ 10 years*	*4 cents/ kwh*	*5 cents/ kwh*	*6 cents/ kwh*
6	1,670	1,333	13,330	$530	$670	$800
8	1,250	1,000	10,000	400	500	600
10	1,000	800	8,000	320	400	480
12	880	664	6,640	266	332	400

How Much Energy Does Your Air Conditioner Use?

To find the energy used by your air conditioning system, you need your electrical bills for the last cooling season and the bills for the month before and after the cooling season. The before-and-after bills should indicate the approximate amount of energy used by all your appliances other than the air conditioner. Find the average amount of energy used during these two months and subtract that amount from the total use during each month you used the air conditioner during the summer. The difference should be the amount of energy used only for air conditioning during each summer month.

But does this value really represent the average amount of energy used by your air conditioner per year? The method is based on the assumption that your electrical consumption for appliances other than the air conditioner doesn't change from month to month and that the only extra electrical consumption during the summer is the air conditioning. While the amount of energy used by any single appliance does change from month to month, on balance the use of one increases when use of another decreases—so the total amount of energy used by all appliances does remain somewhat constant from month to month. In addition to monthly changes, the energy used by some appliances may also change from season to season; but again, some appliances use less during one season while others use more. The energy used for lighting is less during the summer, but the energy used by a refrigerator or freezer is more because of the higher room temperature—so the total electrical consumption remains about the same from season to season.

During the summer, however, you may use some appliances that you don't use during the rest of the year: outdoor electric grill, lawn tools, and dehumidifier. Of these appliances, the dehumidifier can represent a large extra electrical load. If you use a dehumidifier during the summer but not during the rest of the year, you should calculate the amount of energy it uses during the summer months and take that value into account when calculating the amount of energy used by your air conditioner. For example, suppose you have both air conditioner and dehumidifier and your average electrical consumption for the months before and after the cooling season is 525 kwh. If your electrical bill for July is

1,075 kwh, your air conditioner and dehumidifier use 550 kwh (1,075 − 525) during July. If you calculate that your dehumidifier uses 125 kwh per month, your air conditioner uses 425 kwh (550 − 125) during July. Use table C-88 (Appendix C) to calculate the amount of energy used by your air conditioner for last year. If you have a dehumidifier, you will need to do some further calculations in the next section of this chapter.

But was last summer an average summer? If the season was warmer or cooler than usual, the amount of energy used for your air conditioning last year would be higher or lower than your actual average. Obtain a more accurate value for your average energy consumption by repeating the calculations in table C-88 (Appendix C) for all the previous years (the last five should be sufficient) for which you have electric bills. Record your results in table C-89 (Appendix C), then average all these values to obtain the average figure for your annual energy consumption.

Conservation Measures

You can save energy in four ways: Insulate your house so that a minimum of heat is conducted inside; keep your house shaded from direct sunlight; use your air conditioner sparingly; and use a high-efficiency air conditioner.

Increase the Thermostat Setting for Your Air Conditioner The recommended inside temperature for a house is 78° F. To find the temperature at which your own family feels comfortable, increase the inside temperature one degree per day until you hear complaints about the heat; then set the thermostat at the previous day's setting and keep it there.

Energy Savings As a rough approximation, for every degree you increase the inside temperature, you reduce the energy needed for air conditioning by about 3 percent. For example, if you increase your inside temperature from 72° to 78° F., your energy consumption for air conditioning should decrease by 18 percent (6° increase × 3 percent per degree). Use table C-90 (Appendix C) for calculating the expected savings if you increase the inside temperature of your house.

Add Insulation to the Ceilings and Walls of Your House Just as insulation holds the heat inside during the winter, it also keeps heat out during the summer. The exact amount of energy saved is difficult to calculate because the outside temperature and solar heating vary greatly throughout the day. But the fraction of heat gained through ceiling and walls is approximately the same for all houses, so we can use some approximations to find the expected energy savings from adding insulation.

Energy Savings You can save about 2.5 percent of air conditioning energy for every inch of insulation added to the ceiling. This approximation applies up to a total of 9 inches of insulation. For example, if you increase the ceiling insulation thickness from 3 to 6 inches, you should expect to save about 8 percent (3 inches × 2.5 percent per inch); and if you increase thickness to 9 inches, you should save 15 percent.

If your walls are not insulated, you should expect to save 20 percent by adding 3-1/2 inches of insulation. For example, if your annual energy use for air conditioning is 1,500 kwh, you would save about 300 kwh by insulating your walls. At 5 cents per kwh, this would be a saving of $15 per year. Use table C-91 (Appendix C) to calculate your expected savings from insulating your ceiling, walls, or both.

Use Storm Windows and Doors Install them for the winter or leave them up all year. While this action will reduce conductive heat gain through the windows and doors, opening the windows on cooler days will be more difficult (see the next conservation measure). If you have a choice of installing storm windows and keeping them closed all summer or being able to open your windows on cooler days (and turning off the air conditioner), you would save energy by not installing storm windows.

Energy Savings If you have storms on all your windows and doors, you can expect to save about 8 percent of the energy consumed for air conditioning. Use table C-92 (Appendix C) to calculate your expected savings.

Cooperate With Nature On days that are not too hot, turn off the air conditioner and open your windows. Some people turn on the air conditioner in early summer and leave it on regardless of outside temperature. Even if the outside temperature is less than the temperature inside, the air conditioner must operate to remove the heat generated by appliances, by people in the house, and from the sun shining through the windows. With open windows, however, the circulating air can remove this heat. Or use a window fan; it can cool the house just as well as an air conditioner, if the outside temperature isn't too hot, for a fraction of the energy and cost.

Don't Try to Air-condition the Entire Outside World The out of doors is big—you can't cool all of it! Keep all doors and windows closed when the air conditioning is operating.

Use a Minimum Number of Appliances Inside the House Every kilowatt hour of electricity used by appliances in the house produces 3,413 Btu of heat that the air conditioning must remove. The amount of energy required to remove this heat depends on the EER of the air conditioner. To remove 3,413 Btu of heat, a unit with an EER of 6 uses 0.6 kwh; a unit with an EER of 10 uses 0.35 kwh. Vent the warm air out of the house whenever possible. A clothes dryer should have its exhaust air vented outside; and if you have an exhaust fan over your stove, use it when cooking.

Keep Moisture Producing Activities at a Minimum Every pound of water vapor added to the inside air condenses on the air conditioner coils and uses 1,000 Btu of cooling energy. Use the vent fan in the bathroom when you take a shower and the vent fan over the stove when you boil water.

Keep Direct Sunlight Out Use white drapes, or line your drapes with a white liner to reflect some of the light that comes in the windows. Better yet, install outside awnings to shade the windows. Closed curtains or blinds reduce the heat input from direct sunlight by half, and awnings reduce it more. The solar input through an east or west window is about twice the amount that comes through a south-facing window, so be sure to close the curtains on appropriate windows when the sunlight hits the east or west side of the house.

Trees Can Help Plant them so they provide plenty of shade for the house. Not only do they shade the house from direct sunlight, but they also cool the surrounding air.

Install an Attic Fan If your roof is exposed to direct sunlight most of the day, consider installing an attic exhaust fan. Attic temperature can rise to almost 200°F. during the day, and a fan can keep that temperature down to the low hundreds and reduce heat flow through the ceiling. Be sure the attic has enough vent openings or the fan will draw air through the ceiling, requiring the air conditioner to operate longer. If you have loose fill insulation in the attic, be sure the moving air doesn't blow it around.

Keep the Air Conditioner Shaded Window unit air conditioners and the outdoor coils of whole-house air conditioners should be shaded. Air conditioners work more efficiently if the outside coils are cooler. But don't block air circulation around the coils.

Don't Fight a Dehumidifier A dehumidifier, like any other electrical appliance, produces heat that the air conditioner must remove. If your air conditioner isn't removing enough moisture from the room, its cooling capacity is probably too large for that room; the air conditioner cools the air quickly, and not enough air flows past the coils to allow for efficient removal of water vapor. Therefore to remove more moisture with an air conditioner, you should use a smaller capacity unit.

If you need moisture removal more than cooling, use a dehumidifier rather than an air conditioner. The dehumidifier is designed to remove moisture and does so more efficiently than an air conditioner.

Planning to Buy a New Air Conditioner? Buy a unit of the proper size for your house that has an EER of at least 9 Btu/wh. To determine the proper size of unit for your house, consult your heating/cooling contractor or use the guides published in *Consumer Reports Buying Guide* or other publications describing air conditioning. Find the EER values of most room air conditioners on the market by consulting the AHAM *Directory of Certified Room Air Conditioners.*[2] Consult your local dealers for EER values on whole-house air conditioners.

Generally 115-volt units have higher EER values than the 205-volt or 230-volt units; if you can't find a large unit with an EER greater than 9, you may want to consider buying two small 115-volt window units rather than one large 230-volt unit.

Energy Savings All you need know to calculate the energy savings is the EER of your present unit, the EER of the unit you plan to purchase, and the number of kilowatt hours presently used per year for air conditioning. The fractional reduction of energy consumption is the difference between the EER values of the new and old units divided by the EER value of the new unit. For example, suppose your old air conditioner has an EER of 6.2; your annual energy consumption for air conditioning is 1,000 kwh; and you purchase a new air conditioner with an EER of 9. The difference between the EER values is 2.8 (9.0 − 6.2), so the fractional reduction of energy use is 0.31 (2.8 ÷ 9.0) or a saving of

31 percent. Thus each year you could expect to save about 310 kwh (0.31 × 1,000 kwh), which would mean a saving of about $15.

Calculate your expected savings in table C-93 (Appendix C).

Dehumidifier

The basic operation of a dehumidifier is exactly the same as that of an air conditioner except the entire unit is inside the house. As room air passes over the cold coils, the air moisture condenses on the coils and drips into a bucket as shown in figure 26. Unlike the air conditioner, however, the hot coils are inside rather than outside the house. The energy released by the condensed water and the energy from the electricity used to run the dehumidifier enter the room air as heat. For each pint of water that condenses, the dehumidifier adds about 1,000 Btu to the air from the condensed water vapor and another 1,500 to 2,500 Btu from the electrical energy.

Relative Efficiencies and Costs

The useful work done by the dehumidifier is the removal of water from the air, and the amount of energy input is the number of kilowatt hours used to remove that water. Let's define the

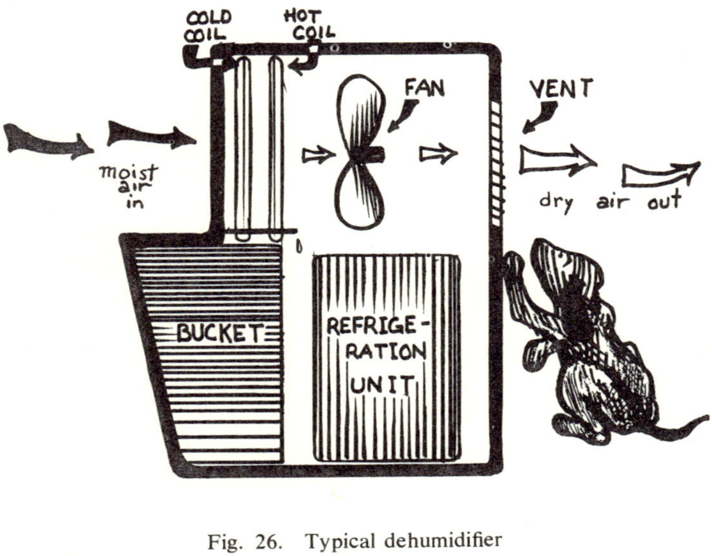

Fig. 26. Typical dehumidifier

efficiency of a dehumidifier as the number of removed pints of water divided by the kilowatt hours of electricity used to remove that water, and call this efficiency the Dehumidifier Energy Efficiency Ratio (DEER).

DEER value = pints of water removed ÷ kilowatt hours used.

To find the DEER value of a dehumidifier, you must know the capacity of the dehumidifier in pints removed in 24 hours and the wattage.

Capacity should be listed on the unit or in the sales information with the unit. The rated capacity of a dehumidifier is the amount of water the unit can remove in 24 hours at a temperature of 80° F. and a relative humidity of 60 percent. At higher temperatures and humidities the capacity is greater, and at lower temperatures and humidities the capacity is less. Wattage, however, may be difficult to find. Many units don't list it on the unit or in the instructions. The amperage, usually listed, is the starting current which is greater than the operating current, so you can't use that to determine the wattage. If the wattage is not listed, use the rotating disc method to find it (see chapter 2).

To find the DEER value of a dehumidifier, divide the capacity of the unit in pints per 24 hours by the number of kilowatt hours used per 24 hours. The number of kilowatt hours used per 24 hours is just the wattage multiplied by 24 hours. For example, if a unit uses 350 watts and can remove 18 pints of water per day, the energy use per 24 hours is 8.4 kwh (350 watts × 24 hours ÷ 1,000) and the DEER is 2.14 pints per kwh (18 pints ÷ 8.4 kwh). Use table C-94 (Appendix C) to calculate the DEER value of your dehumidifier.

Most dehumidifiers have DEER values between 1.9 and 2.5. If your dehumidifier has a DEER value close to 1.9 and is getting old, you may want to consider replacing it with a new unit that has a higher DEER value.

How Much Energy Does Your Dehumidifier Use?

To calculate the amount of energy used by your dehumidifier, you must know the wattage of the unit and its on-time during a season (see Wattage-time Method below); or you must know the DEER value and the number of removed pints of water per season (see Water Brigade Method below). If your dehumidifier has a humidistat and turns on and off periodically, either of these methods will give a fair indication of the amount of energy used; but if your dehumidifier runs constantly, the Wattage-time Method will be better than the Water Brigade Method.

Wattage-time Method Estimate the fraction of dehumidifier on-time during each month of use. Record these fractions in the appropriate blanks of table C-95 (Appendix C) and calculate the amount of energy used each month. If the dehumidifier runs constantly, write 1.0 in the fraction column. The amount of energy used per month is the fraction of dehumidifier on-time multiplied by the wattage and number of hours in one month. As an example of the calculation, a 350-watt dehumidifier that is on 75 percent of the time uses 192 kwh per month (0.75 × 350 watts × 732 hours per month ÷ 1,000 = 192 kwh).

Water Brigade Method If you're unsure of the percentage of dehumidifier on-time each month, you can estimate the amount of energy used by keeping a record of the amount of water removed by your dehumidifier. With this method, unfortunately, you can't calculate the amount of energy used in past years because the information has been unceremoniously poured down the drain—so this method is only good for finding the amount of energy used during the next "drying" season.

Keep a monthly record of the pints of water that the dehumidifier removes from the air. Divide this amount by the DEER value of your dehumidifier to find the number of kilowatt hours used to remove this water. Accuracy of this method depends on how near the room air is to the standard values of 80° F. and 60 percent relative humidity. If the temperature and humidity in your house are above these values, the calculated energy will be more than the actual energy used; if temperature and humidity are below these values, the calculated amount will be less than the actual amount used. You can make this method more accurate by measuring the actual wattage and the actual 24-hour capacity of your unit under its normal operating conditions, then calculating the "operating" DEER value from these values.

Measure the water by marking pint levels on the side of the bucket so you don't have to pour the water into a measuring container each time you empty the bucket. To calibrate the bucket, pour water into the bucket one pint at a time, mark the level for each pint, and label each mark. Make a chart and post it near the dehumidifier or wherever you empty the water so that you can record the amount each time you empty the water. Add the amount of water removed during each month and record that monthly total in table C-96 (Appendix C), then calculate the amount of energy used.

Conservation Measures

All you can do to reduce the amount of energy needed to dehumidify a room is keep as much moisture as possible out of the room and use a high-efficiency dehumidifier.

Seal the Room as Much as Possible Moisture enters as water vapor in the infiltrating air or as water seeping through basement walls and floor. Seal all walls and floors below ground and seal around all windows and doors. In basements, seal the walls with a waterproofing paint, and seal all cracks in the floor.

You Can't Dry Out the Entire Outside World Keep all doors and windows closed when the dehumidifier is on. Air entering an open door or window almost instantly replaces the moisture removed by the dehumidifier.

Use Vent Fans These will vent excess moisture from cooking and bathing areas. Do not vent a clothes dryer inside the house. In general, keep moisture producing activities to a minimum in rooms that are being dehumidified.

Check the Humidity of Your Room If your dehumidifier runs constantly, is it really doing any good? Purchase a hygrometer (humidity gauge—a desk-type temperature-humidity indicator is adequate), turn the dehumidifier off for a few hours, and check the humidity of the air. Then turn on the dehumidifier again and check the humidity 2 or 3 hours later. If the room humidity is only slightly less with the dehumidifier on than off, the dehumidifier is doing little or no good. Moisture is entering the room as fast as the dehumidifier can remove it, so you need to decrease the flow of moisture (preferably) or buy a higher capacity dehumidifier.

Don't Try to Make the Room into a Desert At a low humidity setting, the dehumidifier must run more to keep the humidity low. Find the maximum humidity that is satisfactory and keep the dehumidifier set to maintain that humidity.

Planning to Purchase a New Dehumidifier? Calculate the DEER value of all the units you are considering. If you can't find the wattage listed and the salesperson doesn't know it, write the company for the information or forget about that model. Don't accept statements from the salesperson like, "Oh, it shouldn't use much," or "I'm sure it's quite small," or "I think it's about — watts." Get a reliable wattage number. A "should-be-small" wattage may turn out to be large compared to other models; you're the one, after all, who will be paying for the electricity.

Energy Savings

All you need know to calculate the energy savings is the DEER value of your present unit, the DEER value of the unit you plan to purchase, and the number of kilowatt hours presently used per year by your dehumidifier. The fractional reduction of energy consumption is the difference between the DEER values of the new and old units divided by the DEER value of the new unit. For example, suppose your old dehumidifier has a DEER value of 2.0; electrical consumption for the dehumidifier is 750 kwh; and you purchase a new dehumidifier with a DEER value of 2.4. The difference between the DEER values is 0.4 (2.4 − 2.0), so the fractional reduction of energy use is 0.17 (0.4 ÷ 2.4) or a saving of 17 percent. Thus you could expect to save about 130 kwh per year (0.17 × 750 kwh), which means a saving of about $6. Use table C-97 (Appendix C) to calculate your expected savings from the purchase of a more efficient dehumidifier.

This calculation assumes, however, that your new dehumidifier is set to maintain the same humidity as your old unit. Because the humidistats on different dehumidifiers have different scales, you should invest in a hygrometer (humidity gauge) and measure the air humidity with the old unit running—then set the humidistat of the new unit to give the same humidity. If you want to maintain a lower humidity than before, your saving will be less than that calculated in table C-97.

If your old unit ran constantly and you also plan to run your new unit constantly, you will save energy only if the wattage of the new unit is less than the wattage of the old. A new unit with higher wattage will use more energy than the old unit; but if the new unit has a higher DEER value, a higher capacity, or both, it will remove more moisture from the air than the old unit.

7.

Laundry Appliances

The energy required for laundering clothes includes the amount for the clothes washer, clothes dryer, and iron. The different models of these appliances (except for some clothes washers) consume about the same amounts of energy for the same amounts of use. Thus the amount of energy used by any particular family to wash, dry, and press clothes depends mostly on how many clothes they wash and whether they use a clothes dryer or iron; a family can get by without a clothes dryer if they use a clothesline—and can do without (or almost without) an iron if they buy permanent-press clothing.

Clothes Washer

Basic operation of a clothes washer is simple. The washer agitates the clothes in soapy water, spins them to remove the water, then rinses and spins them again. The only complicated part of the operation is the timer, which turns the water on and off, starts and stops the motor and tub spin, and so on. The timer essentially replaces the human brain; you don't have to think about pouring water or turning switches.

A clothes washer uses energy in two forms: electricity to run the motor and hot water. The motor which turns the agitator and spins the tub to dry the clothes uses about 0.25 kwh per load of wash—and for a cold wash, this is the entire amount of energy used by the washer. But any hot or warm water used requires energy to heat the water.

The amount of hot water used depends on the temperature setting for the wash and rinse water and the water capacity of the tub. Most late model washers have settings for about every combination of hot, warm, and cold wash and rinse-water temperatures: hot/warm; hot/cold; warm/warm; warm/cold; and cold/cold. The water for a hot wash is all hot, but water for a warm wash (or rinse) is half hot and half cold, so the amount of hot water for a warm wash is half the total water in the tub.

How Much Energy Does Your Clothes Washer Use?

To find the amount of energy you use for your laundry, you must know the amount of electrical energy required to run the motor and the amount of hot water used for each load. First, let's consider the electrical energy requirements. The most accurate way to find this amount is to measure it. Use the rotating disc method (see chapter 2) to measure the wattage during an agitation or spin cycle; then find the total time that the clothes washer is agitating and spinning, either in the instructions or by measuring the time yourself during a wash. Record these values in table C-98 (Appendix C). You can ignore the small amount of energy needed to operate the water valves and timer.

If you don't want to bother with measuring the wattage and time, you can use the approximation that most washer motors use about 0.25 kwh per load of laundry. Your washer may use slightly more or less than this amount, but the difference should be small.

Now what about the hot water? If you have the instructions for your clothes washer, read the next section below. Skip to one of the following sections if you don't have the instructions or if your instructions don't list the water used per fill.

Instructions Available This method is the easiest way to find your hot water use and is accurate enough for our purposes. The instructions should list the water used per fill for a regular-size load as well as for any possible smaller loads. Clothes washers fill with water until a preset volume is in the tub, so the amount of fill doesn't depend on water flow or pressure. For a hot/cold setting, the amount of hot water used is the amount listed for one fill—the wash is all hot water and the rinse is all cold water. For a hot/warm setting, the amount of hot water used is one and one-half times the amount listed for a fill—one filling of hot water for the wash and half of one fill for the warm rinse (half hot water, half cold water). The amount of hot water used for a warm/cold setting is half the amount used for a fill (half hot water for the wash, no hot water for the cold rinse); and for a warm/warm setting, the amount of hot water used is the amount listed for a fill (half hot water for the wash, half hot water for the rinse).

Some clothes washers have a short spray rinse during the spin that follows the wash cycle. The temperature of this rinse water is the same as the temperature of the rinse—warm for a warm rinse and cold for a cold rinse. The spray rinse operates for a set time, so the amount of water used depends on two factors: the on-time for the spray rinse and the rate of water flow. The on-time is easy, but flow is not practical, to measure. You can measure the total amount of water used for the spray rinse (and the Water Brigade Method given below instructs how to do this); but that operation is a little messy, so let's take an easy way out and assume that the spray rinse uses 6 gallons of water—a value that represents an average amount used by several clothes washers. Thus a warm spray rinse uses 3 gallons of hot water (half the full amount), and a cold spray rinse uses none. From your instructions, record the data for your clothes washer in table C-99 (Appendix C) for the load sizes and temperature settings you use.

Consumer Reports *Available* Consult the *Consumer Reports Buying Guide* or the monthly issue of *Consumer Reports* that lists hot-water use of the various clothes washers tested. The hot-water amounts for the washers tested are more accurate than the values you find with the preceding section but *Consumer Reports* doesn't test all models and settings. Also, it tests only the models currently on the market, so your washer won't be included if it's no longer being sold. If *Consumer Reports* does list your washer capacity, record the amount of hot-water use in table C-99 (Appendix C).

Water Brigade Method If you can't use the previous methods because you have no instructions or because your model is not listed in *Consumer Reports,* there is yet another way to find the amount of hot water used. (Fetch your sailboat and rubber duckie for this one—and maybe a mop, bucket, and raincoat.) If your washer empties into a laundry tub you're all ready; but if not, you'll need a large tub or plastic trash can—one large enough to hold the water in the washer. Check to be sure the "receiver" is larger than the "sender" by measuring the diameter and height of both the washer tub and the can or tub into which the water will empty.

Empty the wash water into the tub and measure the water volume. Use the dip-pour method to measure the amount of water—dip out the water with a gallon container, pour it down the drain, and count the number of gallons emptied. If your washer has a spray rinse, turn off the washer before the spray rinse comes on so you can measure the spray rinse separately. Then measure the amount of water that emerges during the spray rinse—half that amount, not the 3 gallons we assumed, is the hot water for the warm spray rinse. Record all your values in table C-99 (Appendix C).

Adding It All Up

Now add up the total amounts of hot water and electrical energy that you use for a month's laundry. For example, suppose your clothes washer uses 30 gallons of hot water for a hot/warm wash and 10 gallons for a warm/cold wash. If you wash eight hot/warm loads and fifteen warm/cold loads per month, you would use 390 gallons of hot water plus about 6 kwh to run the motor per month. The hot water would cost you about $1.75 for a gas or oil hot-water heater; and about $3.50 for an electric hot-water heater. Adding the cost of electricity to run the motor, your monthly energy cost for washing clothes would be about $2 for a gas or oil hot-water heater, or $4 for an electric hot-water heater. Use table C-100 (Appendix C) to calculate your monthly costs.

Conservation Measures

The easiest way to save energy is to not do any laundry, but that method doesn't usually win friends and influence people (or may influence them the wrong way!). Because most of the energy used by a clothes washer is in the form of hot water, try to use as little hot water as possible. The following suggestions may help.

Always Wash a Full Load of Laundry—No More, No Less A

washer uses the same amount of energy for a half-load as for a full load, and overloading your washer may damage it. If your washer has settings for medium or small-size loads, use them for your smaller loads.

Use a Cold-water Wash Whenever Possible To do this, you must be selective about the fabrics you buy. But always use hot water for underclothing, for laundry from a family member with a communicable disease, and for diapers.

Use Cold Rinse Water Whenever Possible If your clothes washer has a hot/warm setting but no hot/cold setting, you can obtain a cold rinse by resetting the wash/rinse control to warm/cold or cold/cold after the wash water has filled—but before the rinse water starts to fill. Reset the controls, if your washer has a spray rinse, before the spray rinse begins.

Energy Savings

The amount of energy saved is just the amount consumed for heating the water that would be used for a warm rinse. If your clothes washer uses 10 gallons of hot water for a warm rinse plus 3 gallons for the spray rinse, you would save 13 gallons of hot water each time you eliminated a warm rinse. Eliminating ten warm rinses per month would save you 130 gallons per month—a cost saving of 40 cents if you have an oil or gas hot-water heater or $1.30 if you have an electric hot-water heater. Use table C-101 (Appendix C) to calculate your expected savings by reducing the number of warm rinses.

Planning to Purchase a New Clothes Washer?

Determine the amount of hot water that each model uses. To compare models, you only need to know the comparative amounts of water per fill—the lower the better. Most washers use between 16 and 25 gallons of water per fill, so try to find a model with the features you want that uses no more than 18 to 20 gallons per fill.

Energy Savings The amount of energy saved is just the total energy required to heat the hot water saved by the new clothes washer. Use table C-102 (Appendix C) to calculate your expected savings.

Clothes Dryer

Now that you have a pile of clean, wet laundry, what are you going to do with it? You could hang it on a clothesline, either inside or outside, or use a clothes dryer. Drying clothes outside requires no energy except from the person hanging up the laundry—the solar energy is free. Hanging clothes inside to dry uses no energy unless the house is air conditioned or heated. During the summer, the air conditioner uses energy to remove the water that evaporates from the clothes; and during the winter, the heating system must provide the heat to evaporate the water. A clothes dryer, of course, uses energy.

An electric dryer has a motor and a heating element; a gas dryer has an electric motor and a gas burner. The motor turns the drum that tumbles the clothes, and the heating element or burner produces the heat that evaporates the water from the clothes. The amount of energy needed to dry clothes in a clothes dryer depends on the amount of water retained by the clothes. Cotton fabrics retain more water than permanent-press fabrics, so cotton fabrics require more energy to dry—a pound of cotton fabrics needs about 1,700 Btu to dry, and a pound of permanent-press fabrics needs about 1,100 Btu.

How Much Energy Does Your Clothes Dryer Use?

You can calculate the amount of energy used by your clothes dryer in one of three ways. The first method uses the approximations just mentioned for the amount required to dry 1 lb. of clothes; the second method requires measurement of the power consumed by your dryer and measurement of dryer on-time; and the third method uses your electric or gas meter to show the total amount of energy used per load of laundry. The first method is the easiest, but the other two are more accurate.

Approximation Method As an approximation, all dryers use 1.7 cubic feet of gas or 0.5 kwh of electricity to dry 1 lb. of cotton fabrics; and use 1.1 cubic feet of gas or 0.3 kwh of electricity to dry 1 lb. of permanent-press fabrics. To determine the amount of energy used to dry your own laundry, weigh a typical load of laundry *before washing it;* the weights used in this approximation are dry laundry weights. You can use your bathroom scales—but because some scales are inaccurate at low ranges, you may want to weigh yourself and the laundry (held in a large plastic bag) together, then find your weight without the laundry. The difference between the weight of you holding the laundry and your weight alone is the weight of the laundry.

To find the amount of energy needed to dry a load of laundry, just multiply the laundry weight by the approximate number of cubic feet of gas or kilowatt hours of electricity required to dry 1 lb. of laundry. An 8-lb. load of cotton fabrics would require 13.6 cubic feet of gas (8 lb. × 1.7 cf per lb.) or 4 kwh of electricity (8 lb. × 0.5 kwh per lb.). An 8-lb. load of permanent-press fabrics would require 8.8 cubic feet of gas (8 lb. × 1.1 cf per lb.) or 2.4 kwh of electricity (8 lb. × 0.3 kwh per lb.). If a load is half cotton fabrics and half permanent press, it would require 11.2 cubic feet of gas (6.8 cf for the cotton fabrics and 4.4 cf for the permanent press), or 3.2 kwh of electricity (2 kwh for the cotton, 1.2 kwh for the permanent press). Use table C-103 (Appendix C) to calculate the amount of energy required to dry your different loads of laundry.

Power-time Method A more accurate method to find the amount of energy needed to dry a load of laundry is to measure the wattage or gas flow of your dryer and the on-time of electricity or gas. This method is satisfactory for dryers that heat at a constant rate during the entire drying time but is unsatisfactory for dryers with sensor control, in which heat and energy consumption decrease as the clothes dry. If your dryer has a sensor control, you can't use this method.

The amount of wattage or gas flow should be listed in the instructions or on the dryer; but if not, determine this value by the rotating disc method (electric) or the test dial method (gas). If your dryer has several temperature settings, be sure you know the wattage or gas flow at each setting.

The time in which the dryer uses electricity or gas to heat the clothes does not necessarily correspond to the time from switch-on to turn-off. The permanent-press drying cycle has a cool-down period at the end of the cycle, generally 3 to 7 minutes. The regular fabric cycle may also have a cool-down time, but probably shorter than for the permanent-press cycle. During this cool-down time, the dryer just tumbles the clothes with heat off. Find the cool-down time either by consulting the instructions or watching your electric or gas meter to see when it stops. The cool-down time is the period from when the electricity or gas stops to when the dryer stops; and the drying time is the total on-time, minus the cool-down time, of the dryer.

The amount of energy required to dry a load of clothes is the amount of wattage or gas flow multiplied by the on-time of the electricity or gas. A 5,000-watt dryer that takes 40 minutes (0.67 hour) to dry a load of laundry uses 3.3 kwh (5,000 × 0.67 ÷ 1,000 wh per kwh). A gas dryer with a flow of 17 cubic feet per hour uses 11.4 cubic feet of gas to dry a load of laundry in 40 minutes (17 × 0.67 = 11.4 cf). Use table C-104 (Appendix C) to calculate the amount of energy required to dry different loads of your laundry.

Read Meters Your most accurate way to find the amount of energy used by your dryer is to read your gas or electric meter before and after drying a load of laundry. The difference between the two readings is the amount of energy used. Be sure, however, that no other appliance using the same form of energy as the dryer is operating during that time. Turn off the refrigerator, freezer, lights, air conditioner, electric hot-water heater and other electrical appliances if you have an electric dryer; and turn off the gas hot-water heater if you have a gas dryer. (Don't turn off the pilot light—just set the temperature control to "off" or at the lowest setting.) If you use this method during the heating season, you must turn off the heat also, so you may want to wait until summer—or a day when the house won't get cold if the heat is off for an hour. Be sure to turn each appliance on again after you're finished.

If you have an electric dryer, read your electric meter just before and after drying each type of load. You must read the meter to the nearest tenth of a kilowatt hour in order to attain some reasonable accuracy with this method (see Appendix B for more details). For example, if the initial meter reading is 4,856.6 and the final reading is 4,860.8, the dryer uses 4.2 kwh (4860.8 − 4856.6) to dry the laundry. Use table C-105 (Appendix C) to find the energy used to dry different types of loads in your electric dryer.

If you have a gas dryer, you must expend a little more effort. A gas dryer uses from 5 to 20 cubic feet of gas to dry a load of clothes, depending on the type of material and the size of the load—but the last dial on the gas meter indicates hundreds of cubic feet. Thus you cannot use that dial to measure the gas used by your dryer; you must baby-sit with the gas meter and count the revolutions of the 2-cubic-foot dial. Note where the needle on the dial is pointing when you turn on the dryer, then record the number of complete revolutions and the last fraction of a revolution during the on-time of the dryer. For example, if the 2-cubic-foot dial makes six complete revolutions and one-half of another, the dryer uses 13 cubic feet of gas (6-1/2 × 2) to dry the laundry. Use table C-106 (Appendix C) to record the number of dial revolutions and to calculate the amount of energy used to dry different types of loads.

Adding It All Up

To find the total amount of energy used each month for drying clothes, you must multiply the amount used to dry each type of load by the number of same load types that you dry each month. For example, suppose you dry seven loads of cotton fabrics, thirteen loads of permanent-press fabrics, and five loads of mixed laundry per month. If the energy per load for these three types is 4 kwh, 2.4 kwh, and 3.2 kwh, respectively, the cotton loads use 28 kwh, the permanent-press loads use 31.2 kwh, and the mixed loads use 16 kwh per month. Your total monthly energy consumption for drying clothes would be 75.2 kwh (28 + 31.2 + 16) at a cost of $3.75.

If you have a gas dryer, the calculation for your monthly energy use is the same, with units expressed in cubic feet rather than kilowatt hours. A gas dryer also uses some electricity to run the motor, however, and you must include that energy as part of the total amount used to dry the clothes. Although you could measure the wattage of the motor and find its on-time in order to calculate the energy it uses per load, you can use the approximate value of 0.15 kwh per load because the amount is so small. Thus if you dry twenty-five loads of laundry per month, the gas dryer would use 3.8 kwh of electricity (25 × 0.15) in addition to the gas. Use table C-107 (Appendix C) to calculate the monthly energy consumption of your clothes dryer.

Conservation Measures

If you wash your clothes, you obviously must dry them too. Hanging clothes outside to dry uses no energy (except some of yours), but most people prefer to use a clothes dryer. If you use a dryer, the following suggestions may help to save some energy.

Don't Dry the Clothes Too Long Find the minimum amount of time required to dry a load of laundry and operate your dryer no longer than that time. If your dryer has an automatic dryer control, don't depend on it. That control generally allows the clothes to overdry, using up to 20 percent more energy than really needed.

Keep the Filter Clean Blocked air flow doesn't allow the dryer to operate properly, causing it to use more energy.

Buy Permanent-press Fabrics They require less drying time and save energy needed to iron them (your energy as well as electrical energy).

Don't Place Soaking Wet Clothes into the Dryer Be sure the clothes you insert are as dry as possible. If you can wring water from the clothes, wring out each piece—every pint you wring out saves about a cubic foot of gas or 0.3 kwh of electricity.

Planning to Purchase a New Gas Dryer? Buy one with a mechanical or electrical starter rather than one with a pilot light.

Energy Savings

The amount of energy saved is the amount that would be used by a pilot light. Use your previously calculated value for pilot-light consumption in table C-108 (Appendix C) to find your expected savings.

Hanging Clothes Inside During the Winter—How Much Do You Really Save?

Each pound of water in wet laundry requires 1,000 Btu of heat from the surrounding air to evaporate. A pound of cotton fabric contains about 0.8 lb. of water when it emerges from the clothes washer, so the heating system must supply about 800 Btu of heat to dry that pound of clothes. Thus you save only 900 Btu (1700 that would be used by a clothes dryer minus 800 from the air) by hanging cotton clothes inside the house during the winter rather than using a clothes dryer.

People who have electric heat and an electric clothes dryer would save about 0.25 kwh per pound of cotton laundry. Those with gas heat and a gas dryer would use 1.1 cubic feet of gas in the furnace to produce 800 Btu of room heat (assuming a furnace efficiency of 70 percent), so would save 0.6 cubic feet of gas per pound of clothes dried. Those with oil or gas heat and an electric dryer would save the 0.5 kwh of electricity that would be used to dry the pound of laundry in the dryer—but the furnace would use an extra 0.01 gallons of oil or 1.1 cubic feet of gas to replace the heat removed by the evaporation from the clothes.

For a typical 8-lb. load of laundry, families with electric heat and dryer would save 2 kwh, or about 10 cents, per load. Families with gas heat and dryer would save 5 cubic feet, or about 1.5 cents, per load. And those with oil or gas heat and electric dryer would save 16 cents on the electricity but spend 5 cents for the extra oil or gas for a net savings of 11 cents. If the clothes require ironing because they dried on a clothesline, however, the amount of energy used by the iron would reduce these savings.

If you increase the humidity in the house with a humidifier, moisture from the drying clothes replaces some of the moisture that the humidifier would normally produce. But the water evaporating from one 8-lb. load of laundry releases two to four times more water vapor to the air than would a humidifier during an equal amount of time. Most of this water vapor is quickly lost by convection; the humidifier must start again after the clothes are dry, so you save only a little energy.

Venting the Dryer Inside

A dryer uses 1700 Btu to dry a pound of cotton laundry, but not all of that energy is available as heat if the vented air stays in the house. Evaporating the water in a pound of cotton laundry requires 800 Btu; and because the resulting water vapor leaves the house by convection, these 800 Btu are lost. Thus only about half the heat is left as useful heat. Except for a house with oil or gas heat and an electric dryer, the savings on heat per load of cotton laundry are the same as for hanging clothes inside to dry. The savings per pound of laundry for homes with oil or gas heat and an electric dryer are just the amount of oil or gas saved that would produce 900 Btu of heat—about 4 cents worth of oil or gas per 8-lb. load of cotton fabrics.

With the dryer vented inside, of course, you need a good filter to prevent lint from escaping the dryer; and during the dryer on-times, humidity in the house increases greatly. If a humidifier is used to increase humidity in the house, moisture from the dryer replaces some of the moisture that the humidifier would produce—but the energy saving is small.

Iron

The days of the iron are numbered. With permanent-press shirts, permanent-press pants, and permanent-press sheets, need for an iron is decreasing every year. For some people, however, the amount of energy used by an iron can be a measurable part of the total energy used for the laundry.

An iron is just a flat piece of metal that flattens out wrinkles by heating the clothes fabric. The process is simple, and the original irons were just flat pieces of iron with a handle and were heated by a fire. Then came electricity, and an electrical heating element was built into the iron.

The operation of an iron is similar to that of a waffle iron, cooking grill, or other kitchen appliance with a thermostat. The iron requires some time to preheat to the desired temperature, and during this time is using electricity constantly. Once the iron reaches the chosen temperature, the electricity turns on and off periodically to maintain that temperature. A steam iron may require additional heating time for each addition of water.

How Much Energy Does Your Iron Use?

To find the amount of energy used by your iron, you must know its wattage, the time needed to preheat the iron, and the fraction of electricity on-time while you are ironing. The wattage should be listed on the iron; if not, use the rotating disc method to find it (see chapter 2). To find the fraction of iron on-time while you are ironing, you need help from someone else, unless you want to station your ironing board in front of your electric meter. While you are ironing, the other person should watch the electric meter disc and measure the on-time (that is, the rotating time of the meter disc) and the off-time. The fraction of on-time is the time the electricity is on during one cycle divided by the time required for one cycle. The total amount of energy used for ironing is the preheat energy (wattage × the preheat time) plus the energy used for the actual ironing (wattage × ironing time × fraction of electricity on-time).

For example, suppose your 1,000-watt iron requires 15 minutes to preheat; during the ironing process, the electricity is on for 2 minutes and off for 1 minute and 15 seconds (1.25 minutes). The cycle time is 3.25 minutes (2 + 1.25), so the fraction of electricity on-time is 0.62 (2 ÷ 3.25). The preheat energy is 0.25 kwh (1,000 × 0.25 ÷ 1,000 wh per kwh). If you use the iron for 2 hours after preheating, the iron uses 1.24 kwh (1000 × 2 × 0.62 ÷ 1,000 wh per kwh). The total amount of energy required for the ironing is 1.5 kwh (1.24 kwh + 0.25 kwh preheat energy). Use table C-109 (Appendix C) to calculate the amount of energy you use for ironing during one month.

Conservation Measures

The only ways to save energy are to reduce the amount of ironing and to be sure you turn the iron off when not using it.

Buy Permanent-press Clothes They require no ironing and less drying.

Don't Leave a Hot Iron Unattended A hot iron that is not being used not only wastes energy but is also a hazard, especially with small children in the house. Try to plan your ironing so that you don't have to do other things during your ironing time.

8.

Lights, Television, and Other Household Appliances

Now we come to the smaller appliances in the house. Although most families don't use much energy for these appliances, some consume a significant amount of energy for lights, television, and other small appliances.

Lights

The average family uses between 60 and 100 kwh per month for lights. The amount of energy used by any one family, however, can vary widely—from 20 or 30 kwh to several hundred kwh. The large differences arise for two reasons: the number and types of light fixtures in the house; and, most importantly, the personal habits of persons using the lights.

The basic operation of an incandescent light is quite simple. Electricity flows through a thin metal filament in a bulb; the filament becomes hot and glows white. In fluorescent bulbs, the electric current passes through a gas rather than a metal filament, and the gas glows and gives off light. A fluorescent bulb requires about 25 percent more power to start than that used during its normal operation; but this additional power is required for only a second or two, so the energy needed to start a fluorescent bulb is an insignificant part of the energy used by the bulb.

Relative Efficiencies and Costs

The efficiency of a light bulb is small. Only about 3 percent of the energy used by an incandescent bulb is converted to light. The other 97 percent ends up as heat, as you find out when you touch a bulb that has been on for awhile. Fluorescent bulbs, little better, are about 20 percent efficient.

Light is measured in units of lumens. Because the power used by a light bulb is measured in watts, we can define the efficiency as the lumens of light output (useful power) divided by the wattage to operate the bulb (input power); so the units are lumens per watt. Some typical values of lumens output and lumens-per-watt efficiencies of various bulbs are listed in table 40. Notice from the data in table 40 that the efficiency of incandescent light bulbs is greater for the higher wattage bulbs. Because of this higher efficiency, a single 60-watt bulb produces about as much light (890 lumens) as four 25-watt bulbs (940 lumens); yet a 60-watt bulb uses 40 percent less energy than the four 25-watt bulbs. A single 100-watt light bulb has the same light output as

TABLE 40
Light Outputs and Efficiencies of Light Bulbs

Wattage	*Lumens*	*Lumens/watt*
Incandescent:		
25	235	9.4
40	455	11.4
60	890	14.5
75	1,175	15.7
100	1,750	17.5
150	2,800	18.7
30/70/100 3-way	275/1,050/1,325	9.2/15.0/13.3
50/100/150 3-way	550/1,500/2,050	11.0/15.0/13.4
100/200/300 3-way	1,300/3,600/4,900	13.0/18.0/16.3
Fluorescent:		
20 cool white	1,300	65.0
40 cool white	3,150	78.8
40 daylight	2,600	65.0
75 cool white	6,300	84.0

SOURCE: Weast, Robert C., ed. *Handbook of Chemistry and Physics, 54th Edition*. Cleveland: CRC Press, 1973.

four 40-watt bulbs, yet uses 38 percent less energy. But the biggest difference in efficiencies is between incandescent and fluorescent bulbs. A 20-watt fluorescent bulb has about the same light output as a 75-watt incandescent bulb but uses only 27 percent of the energy.

Also note that the light output at the highest wattage of a 3-way bulb is less than that of the same wattage single-filament bulb. A 3-way bulb has two filaments of different wattages: a 50/100/150-watt three-way bulb has a 50-watt filament and a 100-watt filament. When you turn on the light, the lower wattage filament (50 watts) comes on first; at the next setting, the higher wattage filament (100 watts) comes on; and both filaments are on together at the highest setting (150 watts). The light output at the highest setting is the sum of the light outputs of the two filaments (550 + 1,500 = 2,050 lumens)—but because the lower wattage filaments have a lower efficiency than higher wattage filaments, the efficiency of both together is less than that of a single-filament 150-watt light bulb (2,800 lumens).

The numbers in table 40 are for standard bulbs. Some incandescent bulbs are made to last longer than standard bulbs, but these long-life bulbs produce about 10 percent less light than the same wattage standard bulb. If you replace a standard bulb with the same wattage long-life bulb and have enough light, the long-life bulb will save some money for the cost of a new bulb—assuming that the cost of the long-life bulb is not too much more than the cost of a standard bulb. But if you need to replace a standard bulb with a higher wattage long-life bulb to have enough light, the extra cost of the energy can nullify your savings for the bulb. For example, a standard 60-watt bulb is expected to last about 1,000 hours, so it uses a total of 60 kwh over its lifetime. If you need a 75-watt long-life bulb (light output of 1,000 lumens) to replace a standard 60-watt bulb (light output of 890 lumens), the long-life bulb uses 75 kwh, or 15 kwh more than that used by the standard bulb, for 1,000 hours of operation. At 5 cents per kwh, this extra 15 kwh costs 75 cents; but a new bulb costs about 60 cents, so the long-life bulb won't save any money. Cost of the extra electricity, in fact, is more than the cost of a new bulb.

A new light bulb may sometime be marketed that fits an incandescent bulb socket, lasts much longer than present bulbs, has a higher efficiency (possibly three times as efficient)—but alas, it will cost considerably more. If such a bulb becomes available, you should calculate the energy saved by the new bulb over its expected lifetime and compare that saving, plus the saving of replacing standard bulbs, to the cost of the new bulb. For example, if a new bulb that uses 20 watts can replace a standard 60-watt bulb and last as long as twenty of the present 60-watt bulbs (about 20,000 hours), the new bulb would use 400 kwh as compared to 1,200 kwh used by the twenty standard bulbs. The 800 kwh and the cost of twenty standard bulbs saved would mean that you could spend up to $50 for the new bulb and still save money. (This calculation is based on 5 cents per kwh and 60 cents per standard bulb.)

How Much Energy Do Your Lights Use?

To find the amount of energy used by your lights, all you need know is the wattage of each bulb and the hours of on-time for each bulb during a month. Record the wattage of each bulb in the house, room by room, in table C-110 (Appendix C). For a multiple-bulb fixture, add the wattage of all bulbs in the fixture together and record the total as the wattage for that fixture. Then estimate the hours of on-time for each fixture during each month of the different seasons: spring, fall, summer, and winter. Estimate these on-times for each fixture as accurately as you can.

If a light is used every day of the month for a specified time per day, just multiply that daily time by 30 to obtain the amount of monthly use. For a light used only a few times per week or month or intermittently during the day, estimate the on-time as best you can. Undoubtedly the largest error in this calculation will be the on-time estimations; most people tend to overestimate this value, so be as accurate as possible.

You could keep a record of the on-times for each light, but that is a tedious job. Also, when you become conscious of the light's use, you can easily change your habits; thus the on-time you measure is not the same as it would be if you were not conscious of the light's use.

You can easily measure the on-time for some lights without a lot of paperwork. Remove the bulb from its socket, insert a socket with an outlet on the side, replace the bulb in the new socket, and plug an electric clock into the outlet (see figure 27). Each time you turn on the light, you also turn on the clock—so

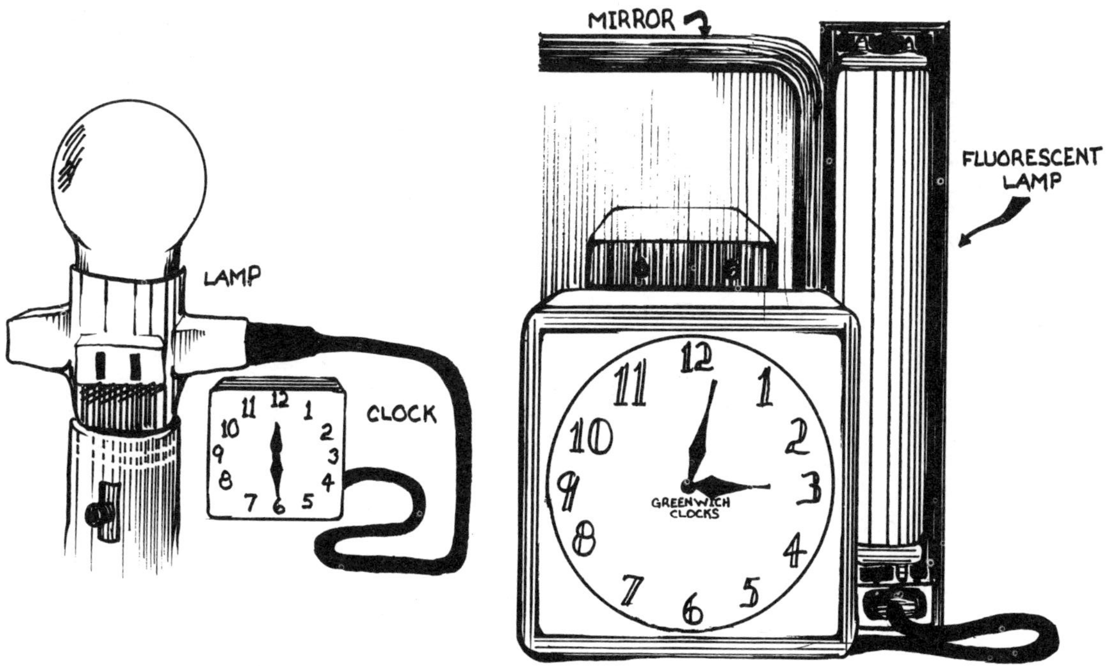

Fig. 27. Lamps with outlet installed for clock

the clock will show the total on-time of the light. Start the clock at 12:00 and periodically check to see when it passes 12:00 again. While this method is fine for table lamps that have space for the extra socket, it's not very practical for ceiling lights unless you want an extension cord draped over the room. If a lamp is plugged into a wall socket controlled by a wall switch, just plug the clock into the same outlet with the light and turn the lamp on and off with the wall switch. Similarly, if a fluorescent bulb is in a fixture controlled by a wall switch and the fixture has an outlet, plug the clock into that outlet.

So now it's time to fetch your pencil and paper and start snoopin' in all your lights to find the number of watts. Record all the wattages and estimated on-times in table C-110 (Appendix C), then calculate the total amount of energy used for your lighting.

Conservation Measures

All you can do to save energy is decrease the use of your lights, decrease the wattage of your light bulbs, or both. If you use more than 60 to 80 kwh per month for lighting (table C-110, Appendix C), you should look for ways to reduce consumption.

During the heating season your actual savings will be less than calculated here, especially if you have electric heat. Refer to chapter 9 for details.

Turn Off Unneeded Lights You may have heard that if you return to a room within 5, 10, or 15 minutes, you can save energy by leaving the lights on rather than turning them off and on again. This is false. Turning the light on and off does shorten the life of a bulb to some extent. In the past when energy was cheap and light bulbs, especially fluorescent bulbs, expensive, you could save money (not energy) by leaving the lights on for short periods rather than by turning them off and on. But energy is no longer cheap, so just form the habit of turning out the lights when you leave a room. Besides, how many times do you intend returning to a room in 5 or 10 minutes but not actually return for several hours?

Energy Savings Refer to table C-110 (Appendix C) and determine if you can shorten any of the on-times for your lights. Record each reduction of time in table C-111 and calculate the expected energy savings. For example, if you reduce the on-time for a 100-watt bulb by half an hour per day, you would save 1.5 kwh per month (100 watts × 15 hrs. ÷ 1,000 wh per kwh).

Don't Light Up a Room More Than Necessary If you are nestled comfortably in a chair, one lamp to light the book is enough—don't illuminate the entire room.

Be Sure That All Light Fixtures Are Clean and Have Clean Reflectors Dust can absorb a lot of light, and a dirty reflector or no reflector at all means that the light directed upward is lost. A clean reflector almost doubles the light output of a bulb.

Use a Clock-timer Rather than leave lights on all the time when you depart the house for several days or weeks, use a clock-timer to turn the lights on and off at the appropriate times. The amount of energy used by the clock-timer is a fraction of that used by a light.

Reduce the Wattage of as Many Lights as Possible Hallways don't have to be bright—a 25-watt bulb is usually sufficient. Many multiple-bulb fixtures put out more light than you really need, so remove as many bulbs from them as possible. Or if you need all the light, replace four or five lower wattage bulbs with two or three higher wattage, more efficient bulbs. For example, five 40-watt bulbs use 200 watts and produce a total of 2,275 lumens of light. Two 75-watt bulbs use only 150 watts and produce 2,350 lumens. But be sure that you don't exceed the maximum wattage bulb specified for that fixture. If you can't use a higher wattage bulb, consider replacing the fixture if the energy saved over several years will be comparable to the cost of the fixture.

Energy Savings The amount of energy saved from changing bulbs in a light fixture is the wattage difference between new and old bulbs multiplied by the on-time of the light fixture. If you replace five 40-watt bulbs in a fixture with two 75-watt bulbs, the wattage difference is 50 (200 − 150). For an on-time of 2 hours per day, you would save 3 kwh and 15 cents per month. Use table C-112 (Appendix C) to calculate your expected savings from reducing the wattage of your light fixtures.

Television

Of all appliances in the home, the most complicated to understand is the television. Basically, the antenna receives the signal from the local television station and this signal goes 'round and 'round inside the set and emerges in front as a picture and sound. In order to make the signal go 'round and 'round and come out a nice picture, energy is needed to operate the picture tube and the transistors or vacuum tubes. We won't detail the 'round and 'round part, but leave that to the television repair people when they "get 'round" to it and send you a bill with lots of round numbers. You should know, however, that the picture tube and smaller vacuum tubes have little heaters that must be hot for the tube to operate—and that these heaters use electricity.

If the set has been off for some time, these heaters are cold and require about a minute to warm up before they can operate. Some manufacturers have come to the rescue of the restless television watcher by developing a set that produces instant picture and sound when the set is turned on. But the on-off switch on the instant-on set doesn't turn the tube heaters on and off—they are on all the time. As long as the set is plugged in, the heaters stay warm, just waiting to jump into action the instant you turn on the set. Of course you pay for this readiness. A vacuum tube instant-on set uses from 50 to 75 watts just to keep the heaters warm, or about 35 to 50 kwh per month; a solid-state set uses from 5 to 40 watts to keep the picture-tube heater warm. An instant-on television has one redeeming quality, however: it does extend the life of the picture tube.

Relative Efficiencies and Costs

Televisions come in many different shapes, sizes, and colors. They all do the same job—produce a picture with sound. The most efficient set is the one that simply uses the least energy to produce the picture. Factors that determine the energy use include the picture size, whether the picture is black and white or color, and whether the set has transistors (solid-state) or vacuum tubes. As a rule of thumb, larger sets use more energy than smaller sets; color sets use more than black and white sets; sets with vacuum tubes use more energy than solid-state sets. The energy range of the different models of same size and type is large, however; so if you're shopping for a new television, you must find the wattage of all the sets you are considering.

How Much Energy Does Your Television Use?

The amount of energy used by your television is just its wattage multiplied by its on-time. If you have an instant-on set, you must also find the off-time wattage. Either consult the instruction manual, the electrical specifications posted on the set, or use the rotating disc method (see chapter 2). For example, if your television uses 150 watts and is on for 6 hours per day, it uses 0.90 kwh per day or 27 kwh per month. Monthly cost for this television is $1.62. Use table C-113 (Appendix C) to calculate your monthly cost.

Conservation Measures

As with lights, all you can do to save energy is to reduce the television on-time. To reduce wattage, you must buy a new set.

During the heating season your actual savings will be less than calculated here, especially if you have electric heat. Refer to chapter 9 for details.

Turn the Television Off When Nobody Is Watching It If you like to listen to the television but don't particularly care to watch the picture, consider buying an FM radio that receives the VHF television channels (channels 2 through 13). That way you can listen to the stations but not use energy to produce a picture.

"Turn Off" the Instant-on Feature Either unplug your instant-on set when not in use or install a switch on the cord to turn the set on and off. Some sets have a switch on the back to turn off the instant-on feature; if your set has such a switch, use it.

Energy Savings The amount of energy saved is just the amount used when the set is off. Use table C-114 (Appendix C) to calculate the energy savings.

Planning to Purchase a New Television? First decide what size and type of set you want, then list the wattages of all the sets you would consider buying. Calculate the cost to operate each set for ten years and add that cost to the purchase price of the set. If the more expensive set uses less energy than a less expensive set, the more expensive set may be cheaper when cost of operation is included.

If you want an instant-on set, purchase one with an "off-wattage" of 15 or less. Even at only 15 watts the set will use about 8 kwh per month, or about 1,000 kwh over the lifetime of the set, for the instant-on feature alone.

Miscellaneous Appliances

Now we come to the other small appliances, each using little energy but which all together can account for a significant part of your total electrical bill. These appliances include electric blankets, clocks, radios, vacuum cleaners, hair dryers, and so forth. Each family owns an assortment of these appliances, and the amount of energy they use depends on the personal habits of each family.

How Much Energy Do Your Appliances Use?

To find the amount of energy used by each appliance, you must know the wattage of the appliance and its on-time. The wattage should be listed on the appliance; if not, measure it by the rotating disc method (see chapter 2). If an appliance has several different settings, such as an electric blanket, either estimate or measure the fraction of maximum wattage for each setting used.

The on-times for some appliances may be hard to figure, but estimate or measure the time as accurately as possible. If your water comes from a well, don't forget the energy required for the water pump. And don't forget the clocks—they may use only 2 to 10 watts, but they are on for 24 hours per day and their energy use can add to a sizable amount. Use table C-115 (Appendix C) for the calculations.

Conservation Measures

You must figure out for yourself just how to reduce your energy use. Be guided by the basic principle: to save energy, just reduce the on-time for any appliance, reduce the wattage if possible—or both. Examine the use of each appliance carefully. Could you use it less or use it more efficiently? Is it on longer than it needs to be? If you do reduce the energy consumption of some appliances, record those savings in table C-116 (Appendix C).

When purchasing any appliance, compare the wattages of several different models to see if they are significantly different; then calculate the expected use over the life of the appliance. Most appliances last about fifteen years.

9.

The Grand Total

Annual Energy Use

Now you come to the final accounting. Record the annual energy use of each appliance in table C-117 (Appendix C). This total is the amount of energy used *before* you take any conservation measures.

You may want to compare the annual total in table C-117 to your actual energy bills, but this comparison is subject to some qualifications. Calculations for the amounts of energy used by the air conditioner and dehumidifier were for average years—but the last year may not have been average. You may have difficulty separating the energy required for heating from the energy used for other appliances, especially if you have electric heat. Also, the energy consumption of a family changes with that family's personal habits; so the energy used by any single appliance changes from month to month and year to year. The totals in table C-117 should come somewhat close to your actual energy bills, but don't be surprised if they don't agree exactly—be happy, in fact, if they come within 20 percent of each other. The main objective of all these calculations is to show you which appliances use little energy and which use a lot and to obtain some approximate values demonstrating this.

When you consider conservation measures, look at the relative amounts of energy use for the different appliances. The ones with highest annual use are the ones that are ripest for reduction. You can also see which ones may be worthless to consider—throwing away an electric can opener that uses 1 kwh per year won't help your total conservation effort much. If you want to total up what you expect to save, use table C-118 (Appendix C) to record the expected savings from the conservation measures that you follow.

When Is a Kilowatt Hour Saved Not a Kilowatt Hour Saved?

Now you have just about reached the end. You have calculated all of the energy used by your appliances, you have carefully decided which conservation measures to follow in order to save energy and money, and you rub your hands in anticipation of all those potential savings. Sorry about this, but here comes a little cold water for your spirits! Pay particular attention if you have electric heat.

Several times in this book we have stated that 1 kwh of electricity produces 3,413 Btu of heat whether you want it or not. During the heating season, these 3,413 Btu per kwh are welcome in the house; that's 3,413 Btu that the heating system doesn't have to produce. So if you reduce the energy consumption of some electrical appliances by one kilowatt hour, that's 3,413 Btu of heat not produced in the house so your heating system must now produce this heat. As an example, suppose you reduce the energy consumption of a certain light fixture by 5 kwh per month, and let's also suppose that you have electric resistance heat. Because the light fixture now produces 5 kwh (17,065 Btu) less heat than before, the heating system now must supply this heat—5 kwh per month. Your total energy saving is a big, fat zero during the heating season. Likewise a television, refrigerator, or any other indoor appliance is a heat source, and each kilowatt hour of reduction in their use results in an additional kilowatt hour used by your heating system. (Some ceiling light fixtures may not contribute 3,413 Btu per kwh to the house because some of that heat may be conducted through the ceiling to the attic.)

But every coin has another side, as well as an edge. Let's consider the edge: During the spring and fall when space heating is not needed, and during times of the summer when you don't use the air conditioner, you save the calculated energy savings. On the other side of the coin: When you use the air conditioner, each kilowatt hour saved by that light fixture actually results in more than a kilowatt hour saved because the air conditioner has less heat to remove from the house.

The same arguments apply if you have gas or oil heat; but because the cost of oil or gas heat is less than electric heat, you do save money by saving electricity during the heating season. If you pay 5 cents per kwh for the "last" kilowatt hour of electricity, the 5 kwh saved per month in our example is a 25-cents saving per month. But your heating system must produce an additional 17,065 Btu (5 kwh × 3,413 Btu/kwh) to replace the 5 kwh not being produced by the light fixture. So a gas furnace must burn an additional 0.24 Ccf of gas at a cost of 7 cents (at 30 cents per Ccf), and the total saving is 18 cents per month. The saving is approximately the same for oil heat. If you have a heat pump for your heating system, each kilowatt hour saved for appliances in the house results in a net saving of about one-half of the 5 kwh. And, of course, every kilowatt hour saved during the cooling season results in a saving for air conditioning.

So if you have electric heat, live it up during the heating season. Watch television until you're bug-eyed, leave lights on as if you're afraid of the dark, open your freezer door and watch the ice melt, or "waste" electricity any way you wish—it won't cost you anything. You must be careful about using hot water and the clothes dryer, however—the hot water from the washer goes down the drain and out of the house along with most of the heat from a clothes dryer, so they contribute little heat to the house. If you have electric heat, the only way to save energy during the heating season is to reduce the inside temperature and do everything possible to reduce heat losses. But when the heat goes off, return to your spartan ways and be careful about your energy use.

If you have gas or oil heat, you can't "waste" energy as can those with electric heat—but be consoled by the fact that you're paying less for heat.

Appendix A
Calculating Areas

The heat loss sections ask you to find the areas of walls, ceilings, doors, windows, and so on. Formulas for finding the area of rectangles, triangles, and circles follow.

Rectangle: Area = Length × Width

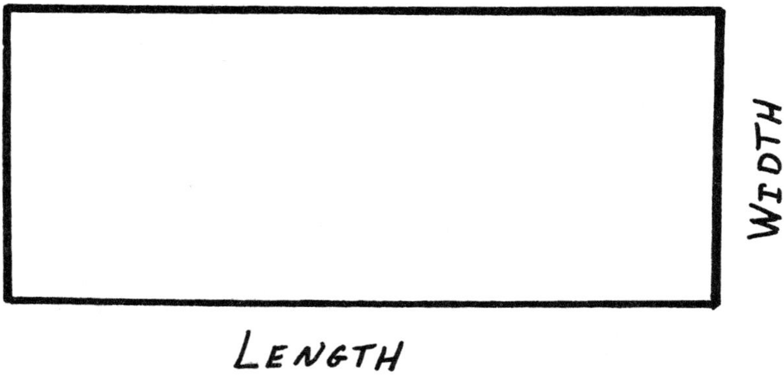

Triangle: Area = 1/2 × Base × Height

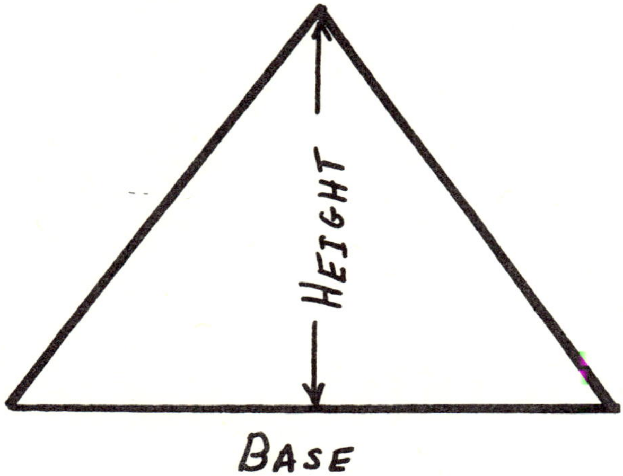

Circle: Area = 1/4 × 3.14 × Diameter squared

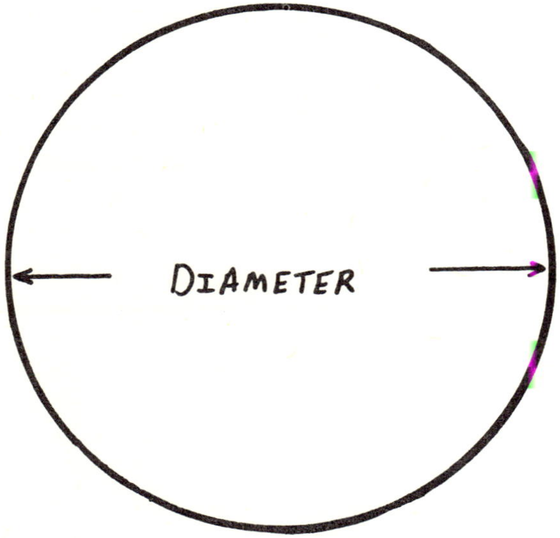

If a surface has a shape that is not a rectangle, triangle, or circle, separate the total area into rectangles and triangles, find the area of each rectangle or triangle, then add those areas together to find the total area. For example, consider the following areas:

The total area is the area of the rectangle plus the area of the triangle.

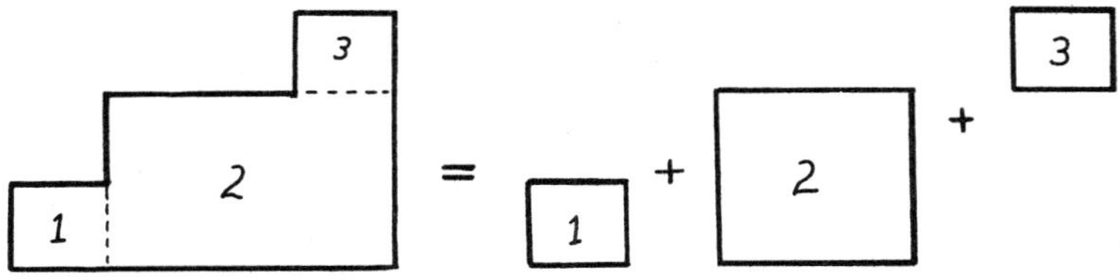

The total area is the sum of the areas of the three rectangles.

Or consider an octagonal window:

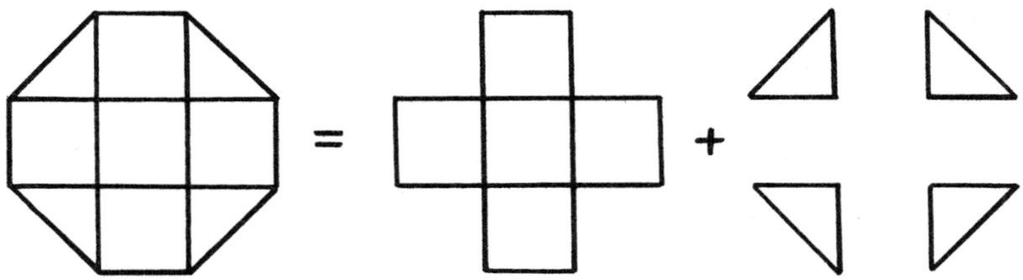

The total area is the area of the five squares (rectangles) plus the area of the four triangles.

Appendix B
Mathematical Tidbits

Significant Figures

If you complete many of the calculations listed in this book, lots of numbers will be flashing before your eyes. For example, divide 37 by 1.7 and your trusty calculator will promptly flash the answer: 21.76470588. But do all those nice numbers really mean anything? No. Only the first three are significant, so the answer is just 21.8 (the .76 is rounded to .8).

With few exceptions, no number calculated in this book is accurate to more than three places. So no matter how many figures your calculator puts out (many extend to ten places), read only three places. If you're confused about how to count places, follow these rules:

a. If the number is greater than 0.1, just count the places from the left: 389572 becomes *390*000, 8432 becomes *8430,* 0.67394 becomes 0.*674*.
b. If the number is less than 0.1, count the places from the first nonzero digit: 0.00375967 becomes 0.00*376*; 0.028314 becomes 0.*0283*.

You can, of course, read all the places and complete all the calculations with big numbers—but be aware, at least, that the numbers have no real meaning beyond the third (or, in some cases, second) place.

Units

Anything that we measure is described by a number and a unit—10 inches, 5 pounds, 15 minutes. Without units, numbers would have no meaning. For example, if someone tells you to talk for a period of 20, is it 20 seconds, 20 minutes, 20 days, or 20 fortnights? In this book we have encountered units of Btu, kilowatt hours, watts, and so on, and these are the units we must consider when calculating the amount of energy used by an appliance. A kilowatt hour (kwh) is defined as 1 kilowatt multiplied by 1 hour; so to find the amount of energy used, we must multiply the wattage in *kilowatts* by a time in *hours*—not minutes, seconds, days, or fortnights. Although we *could* define a kilowatt minute or kilowatt fortnight, the kilowatt hour is the standard measure, so that's the one we use in this book.

Similarly, when multiplying Btu per hour by a certain time to find the number of Btu used, we must multiply by that time in hours. If you consider units as fractions, you can cancel them:

$$\frac{\text{Btu}}{\cancel{\text{hour}}} \times \cancel{\text{hour}} = \text{Btu} \quad \text{(cancel the hours)}$$

The proper units are shown in most of the tables, and you can see how the final units result from the units multiplied or divided in the calculation. In some tables, however, the units are omitted to save space, but you should be able to insert them if you want. For example, the units of the R-value are

$$\frac{\text{hours} \times \text{ft}^2 \times {}^\circ\text{F.}}{\text{Btu}}$$

so that in a calculation for heat loss, the units are expressed in Btu:

$$\text{Heat loss} = \text{Area (ft}^2) \times \text{Ddh (hours} \times {}^\circ\text{F.)} \div \text{R-value} \left(\frac{\text{hours} \times \text{ft}^2 \times {}^\circ\text{F.}}{\text{Btu}} \right) = \text{Btu}$$

Interpolation of Meters

Interpolation is the process of finding a value between two other values. For some calculations involving an electric meter, you may need to find the tenths of a kilowatt hour. For example, if the pointer is halfway between the three and four on the 1-kwh dial, that dial reads 3.5 kwh. If the pointer is about two tenths of the way between the three and four, the meter reads 3.2. Visually divide the space between the two numbers into ten equal divisions and estimate the division at which the pointer rests. For gas meters, the last dial is a measure of hundreds of cubic feet, so if the pointer is halfway between the three and four, the meter reads 3.5 hundreds, or 350 cubic feet, or 3.5 Ccf.

Generally the space between the numbers is small, and the end of the pointer is not sharp. Even if you can't interpolate to a precise tenth, any fractional estimation would be better than reading to the nearest whole number. If the pointer is between the five and six and is close to six tenths, seven tenths, or eight tenths but you aren't sure of its precise position, simply assume one of these values; any choice would be more accurate than rounding to six.

Appendix C
Calculation Tables

Use the following tables to calculate the amounts of energy used by your appliances and the expected savings from applying the suggested conservation measures. Because many of these tables are duplicate methods of calculation or refer to different fuels, you will not use all of them—you'll probably need only about half of them. Don't forget to read only three significant figures in the calculations.

TABLE C-1
Kh Value for Your Electric Meter

Kh value = ______ watt hour per revolution

TABLE C-2
Gas Consumed by Your Pilot Lights

Gas consumed per hour	= 30 ÷ ______	minutes for one revolution of half-cubic-foot dial	= ______	cfh
Gas consumed per month	= ______	cfh × 732 hours per month	= ______	cf per month
Gas consumed by each pilot light	= ______	cf consumed per month ÷ ______	number of pilot lights = ______	cf per month

TABLE C-3
Your Cost of Furnace Oil

Cost = ______ cents per gallon

TABLE C-4
Your Electrical Rate Schedule

Record the categories in column A and the corresponding rates for each category in column B. Record the fuel adjustment charge (FAC) in each blank in column C. Add that value to the basic cost per kwh in column B to find the total cost per kwh in column D. If your utility lists the costs in cents per kwh, convert cents to dollars by dividing the cost in cents by 100 (3 cents per kwh ÷ 100 = $0.03 per kwh).

(A)[a]	(B)[a]	(C)	(D)
0 – ______ kwh:	$______	+ $______ FAC[b]	= $______
or			
Customer charge $______			
next ______ kwh:	$______ per kwh	+ $______ FAC[c]	= $______ per kwh
next ______ kwh:	$______ per kwh	+ $______ FAC	= $______ per kwh
next ______ kwh:	$______ per kwh	+ $______ FAC	= $______ per kwh
next ______ kwh:	$______ per kwh	+ $______ FAC	= $______ per kwh
over ______ kwh:	$______ per kwh	+ $______ FAC	= $______ per kwh

[a]Although six blanks are available, use only as many as you need for your rate schedule. Some utilities have six separate categories, while others have only three or four. If you have lifeline rates, the lifeline quantities will determine some of these categories—see table 5 in text for examples.

[b] Multiply the FAC by the number of kilowatt hours in this category (first number in column A) to obtain the FAC for this category.

[c] For this blank and all blanks below in this column, record the FAC in dollars per kwh.

TABLE C-5
Your Gas Rate Schedule

0 – ______ cf or Ccf[a]	$______		+ $______ FAC[b]	= $______
or Service charge	$______			
next ______ cf or Ccf or therm	$______	per Ccf	+ $______ FAC[c]	= $______ per Ccf or therm
next ______ cf or Ccf or therm	$______	per Ccf or therm	+ $______ FAC	= $______ per Ccf or therm

next _____	cf or Ccf or therm	$_____ per Ccf or therm	+ $_____ FAC	= $_____ per Ccf or therm
over _____	cf or Ccf or therm	$_____ per Ccf or therm	+ $_____ FAC	= $_____ per Ccf or therm

[a]Although five blanks are available, use only as many as you need for your rate schedule. If you have lifeline rates, the lifeline quantities will determine some of these categories—see table 10 in text for examples.

[b] Multiply the FAC by the number of Ccf (or therms) in this category to obtain the FAC for this category.

[c] For this blank and all blanks below in this column, record the FAC in dollars per Ccf (or therm).

TABLE C-6
Your Average and "Last" Electric and Gas Costs

Electric Costs:

a. No Electric Heat

Use your latest bill from either March, April, or November. Divide the total bill by the number of kilowatt hours for the month and record that value in the blank below for the average cost. If your bill includes either state or local taxes or both, don't exclude them—they are a part of your costs. For example, if your bill is $30.97 for 650 kwh, the average cost is $30.97 ÷ 650 kwh = $0.048 (4.8 cents) per kwh.

Refer back to table C-4 to find the cost of the "last" kilowatt hour and record that value in the blank below for the "last" cost. Using the example rate schedule in text table 2, for example, if your consumption was 650 kwh, the last kilowatt hour (650th) would be in the "next-500-kwh" (500–1,000 kwh) category, and the cost would be $0.037 (3.7 cents) per kwh.

b. With Electric Heat

Follow the above procedures, except do the calculations for two months—one month during the heating season (preferably December, January, or February) and one month during the nonheating season (preferably May, September, or October). Record the average and "last" costs in the appropriate blanks below, add the two monthly costs together, and divide the total by two to find the average and "last" costs.

Gas Costs:

If you do not have gas heat, follow the procedures in section a for electric bills. If you have gas heat, follow the procedures in section b.

NOTE: Because utility rates are steadily increasing in many localities, be sure to use your latest bills.

Electric, No Electric Heat:

Average cost = $_____ per kwh
"Last" cost = $_____ per kwh

Electric, With Electric Heat:

	Heating month		*Non-heating month*		*Total*			
Average:	$______	+	$______	=	$______	÷ 2 =	$______	per kwh
"Last":	$______	+	$______	=	$______	÷ 2 =	$______	per kwh

Gas, No Gas Heat:

Average cost = $______ per Ccf
"Last" cost = $______ per Ccf

Gas, With Gas Heat:

	Heating month		*Non-heating month*		*Total*			
Average:	$______	+	$______	=	$______	÷ 2 =	$______	per Ccf
"Last":	$______	+	$______	=	$______	÷ 2 =	$______	per Ccf

Oil:

Average cost = "last" cost = cost per gallon (table C-3) = $______

TABLE C-7
Efficiency and BTU Output of Your Furnace

Gas:

Input[a] = ______ Btu per hour

or:

Input[a] = ______ cf input per hour × 1000 Btu per cf = ______ Btu per hour

OR

Input[a] = ______ Ccf input per hour × 100,000 Btu per Ccf = ______ Btu per hour

Output = ______ Btu per hour output or at bonnet

Efficiency = ______ Btu per hour output ÷ ______ Btu per hour input = 0.______

Btu per Ccf useful heat = 100,000 Btu per Ccf × 0.______ efficiency = ______ Btu per Ccf useful heat

Oil:[b]

Input[a] = ______ Btu per hour

OR

Input[a] =	______	gallon per hour input	×	140,000 Btu per gallon	= ______ Btu per hour

Output = ______ Btu per hour

Efficiency =	______	Btu per hour output	÷ ______	Btu per hour input	= 0.______

Btu per gallon useful heat	= 140,000 Btu per gallon	× 0.______	efficiency =	______ Btu per gallon useful heat

---

[a] Use the blank that corresponds to the units listed for the input to your furnace. If necessary, convert that value to Btu per hour as indicated.

[b] The actual efficiency of your furnace depends on the actual oil and air flow settings of the furnace. If you have your oil furnace cleaned and serviced once a year, ask the service person to find the actual efficiency of your furnace and record that value in the blank for efficiency. This value is the maximum efficiency, however; during the heating season, the efficiency decreases as the burner becomes dirty or the air or oil flow settings come out of adjustment.

TABLE C-8
Energy Used to Heat Your House

Energy used per month during nonheating season[a] = ______ {Ccf / gallon / kwh}

Energy used by other appliances during heating season	= ______ {Ccf / gallon / kwh}	used per month during nonheating season	× ______	months of heating season

= ______ {Ccf / gallon / kwh}

Energy used for heat per year	= ______ {Ccf / gallon / kwh}	total energy used during heating season	− ______ {Ccf / gallon / kwh}	used by other appliances during heating season

= ______ {Ccf / gallon / kwh}

Cost for heat per year = ______ {Ccf / gallon / kwh} used for heat per year × $______ (average rate, table C-6) per {Ccf / gallon / kwh} = $______

[a] Either use any bill from the nonheating months (but not the highest or lowest bill) or find the average of all your bills. If you have oil heat and hot water and if one delivery spans the entire nonheating season plus part of the heating season, try to estimate the amount used only during the nonheating season, and estimate the use per month. If you use oil only for heat, record that value for the number of gallons used for heat per year. If you have electric heat and air conditioning, use only months the air conditioning is not used.

TABLE C-9
Your Daily Average Inside Temperature

Day: ______ °F. daytime temperature × ______ hours per day at day temperature = ______

Night: ______ °F. night temperature × ______ hours per day at night temperature = ______

Other: ______ °F. other temperature × ______ hours per day at other temperature = ______

Total = ______

Average inside temperature = ______ total ÷ 24 = ______ °F. average

TABLE C-10
Your Degree-difference-hours for a Heating Season

Degree-days = ______

Heating days = ______ days; average outside temperature = ______ °F.

Temperature difference = ______ °F. inside temperature (table C-9) − ______ °F. average outside temperature = ______ °F.

Degree-difference-hours = ______ °F. temperature difference × ______ heating days × 24 hours per day = ______ Ddh

TABLE C-11
Conductive Heat Losses Through Your Windows

Glazing: ______ Material: ____________

Heat loss per year = ______ ft^2 area × ______ Ddh (table C-10) ÷ ______ R-value = ______ Btu

Glazing: ______ Material: ____________

Heat loss per year = ______ ft^2 area × ______ Ddh ÷ ______ R-value = ______ Btu

Glazing: ______ Material: ____________

Heat loss per year = ______ ft^2 area × ______ Ddh ÷ ______ R-value = ______ Btu

Total Btu heat loss through windows = ______ Btu

Energy Required to Replace Window Heat Loss:

Gas:

Ccf = ______ Btu heat loss ÷ ______ Btu per Ccf useful heat (table C-7) = ______ Ccf

Oil:

gallons = ______ Btu heat loss ÷ ______ Btu per gallon useful heat (table C-7) = ______ gallon

Electricity:

kwh = ______ Btu heat loss ÷ 3,413 Btu per kwh = ______ kwh

Cost per year = ______ {Ccf / gallon / kwh} × $______ per {Ccf / gallon / kwh} average value (table C-6) = $______

TABLE C-12
Conductive Heat Losses Through Your Outside Doors

Door: ______________________

Heat loss per year = ______ ft^2 area × ______ Ddh (table C-10) ÷ ______ R-value (text table 17) = ______ Btu

Door: ______________________

Heat loss per year = ______ ft^2 area × ______ Ddh ÷ ______ R-value = ______ Btu

Door: ______________________

Heat loss per year = ______ ft^2 area × ______ Ddh ÷ ______ R-value = ______ Btu

Door: ______________________

Heat loss per year = ______ ft^2 area × ______ Ddh ÷ ______ R-value = ______ Btu

Total heat loss through all doors = ______ Btu

Energy Required to Replace Door Heat Loss:

Gas:

Ccf = ______ Btu heat loss ÷ ______ Btu per Ccf useful heat (table C-7) = ______ Ccf

Oil:

Gallons = ______ Btu heat loss ÷ ______ Btu per gallon useful heat (table C-7) = ______ gallon

Electricity:

Kwh = ______ Btu heat loss ÷ 3,413 Btu per kwh = ______ kwh

Cost per year = ______ {Ccf / gallon / kwh} × $______ per {Ccf / gallon / kwh} average value (table C-6) = $______

TABLE C-13
R-values for Your Walls

Wall: ______

R = ______ R-value (inside) + ______ R-value (support) + ______ R-value (outside) = ______

Wall: ______

R = ______ R-value (inside) + ______ R-value (support) + ______ R-value (outside) = ______

Wall: ______

R = ______ R-value (inside) + ______ R-value (support) + ______ R-value (outside) = ______

TABLE C-14
Conductive Heat Losses Through Your Walls

Wall construction: ______

Heat loss per year = ______ ft²[a] area × ______ Ddh (table C-10) ÷ ______ R-value (table C-13) = ______ Btu

Wall construction: ______

Heat loss per year = ______ ft² area × ______ Ddh ÷ ______ R-value = ______ Btu

Wall construction: ______

Heat loss per year = ______ ft² area × ______ Ddh ÷ ______ R-value = ______ Btu

Total heat loss through walls = ______ Btu

Energy Required to Replace Wall Heat Loss:

Gas:

Ccf = ______ Btu heat loss ÷ ______ Btu per Ccf useful heat (table C-7) = ______ Ccf

Oil:

gallons = ______ Btu heat loss ÷ ______ Btu per gallon useful heat (table C-7) = ______ gallon

Electricity:

kwh = ______ Btu heat loss ÷ 3,413 Btu per kwh = ______ kwh

Cost per year = ______ {Ccf / gallon / kwh} × $______ per {Ccf / gallon / kwh} average value = $______ (table C-6)

[a] Be sure to include only the wall area. Deduct the areas of windows and doors from the total side area of the house.

TABLE C-15
R-values for Your Ceilings

Ceiling construction: ______________________

R-value = ______ R-value (inside) + ______ R-value (insulation) + ______ R-value[a] (roof) = ______

Ceiling construction: ______________________

R-value = ______ R-value (inside) + ______ R-value (insulation) + ______ R-value[a] (roof) = ______

Ceiling construction: ______________________

R-value = ______ R-value (inside) + ______ R-value (insulation) + ______ R-value[a] (roof) = ______

[a] Include roof only if ceiling and roof are one structure; that is, no attic above the ceiling.

TABLE C-16
Degree-difference-hours (Ddh) for Your Ceiling (Ceilings below an unheated attic only)

Average attic temperature = ______ °F. outside temperature (table C-10) + ______ °F. temperature difference (table 20, text) = ______ °F.

Temperature difference = ______ °F. inside temperature (table C-9) − ______ °F. average attic temperature = ______ °F.

Ddh for ceiling = ______ °F. temperature difference × ______ heating days (table C-10) × 24 = ______ Ddh

TABLE C-17
Conductive Heat Losses Through Your Ceilings

Ceiling: ______________________

Heat loss per year = ______ ft^2 area × ______ Ddh[a] ÷ ______ R-value (table C-15) = ______ Btu

Ceiling: ______________________

Heat loss per year = ______ ft^2 area × ______ Ddh[a] ÷ ______ R-value = ______ Btu

Ceiling: ______________________

Heat loss per year = ______ ft^2 area × ______ Ddh[a] ÷ ______ R-value = ______ Btu

Total heat loss from ceilings = ______ Btu

Energy Required to Replace Ceiling Heat Loss:

Gas:

Ccf = ______ Btu heat loss ÷ ______ Btu per Ccf useful heat (table C-7) = ______ Ccf

Oil:

gallons = ______ Btu heat loss ÷ ______ Btu per gallon useful heat (table C-7) = ______ gallon

Electricity:

kwh = ______ Btu heat loss ÷ 3,413 Btu per kwh = ______ kwh

Cost per year = ______ {Ccf / gallon / kwh} × $______ per {Ccf / gallon / kwh} average value (table C-6) = $______

[a] Use the degree-difference-hours from table C-16 if the ceiling faces an unheated attic or from table C-10 if the ceiling and roof are a single structure.

TABLE C-18
Degree-difference-hours (Ddh) for Your Basement

Below ground:

Temperature difference	=	______	°F. basement temperature	−	______	°F. ground temperature (table 21, text)	= ______	°F.

Ddh	=	______	°F. temperature difference	×	______	heating days (table C-10)	× 24 hours per day	= ______	Ddh

Above ground:[a]

Temperature difference	=	______	°F. basement temperature	−	______	°F. outside temperature (table C-10)	= ______	°F.

Ddh	=	______	°F. temperature difference	×	______	heating days (table C-10)	× 24	= ______	Ddh

[a] If the basement temperature is the same as the rest of the house, the degree-difference-hour value is the same as that calculated in table C-10.

TABLE C-19
Conductive Heat Losses from Your Basement

Wall, below ground:

Heat loss	=	______	ft^2 area	×	______	Ddh (table C-18)	÷ ______	R-value (table 22, text)	= ______	Btu
Heat loss	=	______	ft^2 area	×	______	Ddh	÷ ______	R-value	= ______	Btu
Heat loss	=	______	ft^2 area	×	______	Ddh	÷ ______	R-value	= ______	Btu

Wall, above ground:

Heat loss	=	______	ft^2 area	×	______	Ddh (table C-18 or C-10)	÷ ______	R-value (table 22, text)	= ______	Btu

Heat loss = ______ ft^2 area × ______ Ddh ÷ ______ R-value = ______ Btu

Heat loss = ______ ft^2 area × ______ Ddh ÷ ______ R-value = ______ Btu

Floor:

Heat loss =	______ ft^2 area ×	______ Ddh ÷	______ R-value =	______ Btu
		(table C-18)	(table 22, text)	

Heat loss = ______ ft^2 area × ______ Ddh ÷ ______ R-value = ______ Btu

Windows:

Heat loss =	______ ft^2 area ×	______ Ddh ÷	______ R-value =	______ Btu
		(table C-18 or C-10)	(table 16, text)	

Door:

Heat loss =	______ ft^2 area ×	______ Ddh ÷	______ R-value =	______ Btu
		(table C-18 or C-10)	(table 17, text)	

Band Joist:[a]

Heat loss =	______ ft^2 area ×	______ Ddh ÷	______ R-value =	______ Btu
		(table C-18 or C-10)	(table 22, text)	

Total heat loss = ______ Btu

Energy Required to Replace Basement Heat Loss:

Gas:

Ccf =	______ Btu heat loss ÷	______ Btu per Ccf useful heat =	______ Ccf
		(table C-7)	

Oil:

Gallons	=	______	Btu heat loss	÷	______	Btu per gallon useful heat (table C-7)	= ______ gallon

Electricity:

Kwh	=	______	Btu heat loss	÷ 3,413 Btu per kwh	= ______ kwh

Cost per year = ______ {Ccf / gallon / kwh} × $______ per {Ccf / gallon / kwh} average value (table C-6) = $______

[a] If you have added insulation to your band joist, add the R-value of that insulation to the R-value for the band joist in text table 22 to obtain your R-value.

TABLE C-20
Degree-difference-hours (Ddh) for the Floor Over Your Crawl Space

Temperature difference = ______ °F. average inside temperature (table C-9) − ______ °F. crawl space temperature (table 24) = ______ °F.

Ddh = ______ °F. temperature difference × ______ heating days (table C-10) × 24 hours per day = ______ Ddh

TABLE C-21
Conductive Heat Losses to Crawl Space Through Your Floor

Floor: ______________________

Heat loss = ______ ft² area × ______ Ddh (table C-20) ÷ ______ R-value (table 23, text) = ______ Btu

Floor: ______________________

Heat loss = ______ ft² area × ______ Ddh ÷ ______ R-value = ______ Btu

Floor: ______________________

Heat loss = ______ ft² area × ______ Ddh ÷ ______ R-value = ______ Btu

Total heat loss through floor = ______ Btu

Energy Required to Replace Floor Heat Losses to Crawl Space:

Gas:

Ccf = ______ Btu heat loss ÷ ______ Btu per Ccf useful heat (table C-7) = ______ Ccf

Oil:

Gallons = ______ Btu heat loss ÷ ______ Btu per gallon useful heat (table C-7) = ______ gallon

Electricity:

Kwh = ______ Btu heat loss ÷ 3,413 Btu per kwh = ______ kwh

Cost per year = ______ {Ccf / gallon / kwh} × $______ per {Ccf / gallon / kwh} average value (table C-6) = $______

TABLE C-22
Conductive Heat Losses Through Your Slab Floor

Heat loss per year = ______ ft perimeter of house × ______ Ddh (table C-10) ÷ ______ R-value (table 25, text) = ______ Btu

Energy Required to Replace Floor Heat Loss:

Gas:

Ccf = ______ Btu heat loss ÷ ______ Btu per Ccf useful heat (table C-7) = ______ Ccf

Oil:

Gallons = ______ Btu heat loss ÷ ______ Btu per gallon useful heat (table C-7) = ______ gallon

Electricity:

Kwh = ______ Btu heat loss ÷ 3,413 Btu per kwh = ______ kwh

Cost per year = ______ {Ccf / gallon / kwh} × $______ per {Ccf / gallon / kwh} average value (table C-6) = $______

TABLE C-23
Infiltration Heat Losses from Your House

Rooms with 0 walls with windows or exterior doors

Room	*Area (ft^2)*
________	________
________	________
________	________
________	________
________	________

Total area = ________ ft^2

Rooms with 1 wall with windows or exterior doors

Room	*Area (ft^2)*
________	________
________	________
________	________
________	________
________	________

Total area = ________ ft^2

Rooms with 2 walls with windows or exterior doors

Room	*Area (ft^2)*
________	________
________	________
________	________
________	________
________	________

Total area = ________ ft^2

Rooms that have 3 walls with windows or exterior doors

Room	*Area (ft^2)*
________	________
________	________
________	________
________	________
________	________

Total area = ________ ft^2

For each of the above groups of rooms, fill in the following blanks and calculate the cubic feet per hour (cfh) infiltration rates.

Number of walls with windows or exterior doors	*Area (ft^2)*		*Wall Height (ft)*		*Air-change factor (table 26, text)*		*Air-change rate (cfh)*
0	________	×	________	×	________	=	________
1	________	×	________	×	________	=	________
2	________	×	________	×	________	=	________
3	________	×	________	×	________	=	________

Total infiltration rate = ________

Heat loss = ______ Cfh infiltration rate × ______ Ddh (table C-10) × 0.018[a] = ______ Btu per year

Energy Required to Replace Infiltration Heat Loss:

Gas:

Ccf = ______ Btu heat loss ÷ ______ Btu per Ccf useful heat (table C-7) = ______ Ccf

Oil:

Gallons = ______ Btu heat loss ÷ ______ Btu per gallon useful heat (table C-7) = ______ gallon

Electricity:

Kwh = ______ Btu heat loss ÷ 3,413 Btu per kwh = ______ kwh

Cost per year = ______ $\left\{\begin{matrix}\text{Ccf}\\ \text{gallon}\\ \text{kwh}\end{matrix}\right\}$ × \$______ per $\left\{\begin{matrix}\text{Ccf}\\ \text{gallon}\\ \text{kwh}\end{matrix}\right\}$ average value (table C-6) = \$______

[a] This value is the number of Btu needed to heat 1 cubic foot of air by 1° F. (see page 58).

TABLE C-24
Hours of Furnace On-Time

Furnace input (gas or oil) = ______ Btu per hour input (table C-7) ÷ $\left\{\begin{matrix}\text{140,000 Btu per gallon oil}\\ \text{100,000 Btu per Ccf gas}\end{matrix}\right\}$ = ______ $\left\{\begin{matrix}\text{gallon}\\ \text{Ccf}\end{matrix}\right\}$ per hour

Furnace input (electric) = ______ watts ÷ 1,000 = ______ kw input

On-time of furnace during heating season = ______ $\left\{\begin{matrix}\text{Ccf}\\ \text{gallon}\\ \text{kwh}\end{matrix}\right\}$ energy used for heat (table C-8) ÷ ______ $\left\{\begin{matrix}\text{Ccf per hour}\\ \text{gallon per hour}\\ \text{kw}\end{matrix}\right\}$

= ______ hours

TABLE C-25
Conductive Heat Losses Through Hot-Air Ducts

Area of Ducts:

Section	*Ft length*		*Ft around*		*Area (ft^2)*
______	______	×	______	=	______
______	______	×	______	=	______
______	______	×	______	=	______
______	______	×	______	=	______
______	______	×	______	=	______

Total area = ______ ft^2

Temperature of hot air = ______ °F.

R-value of insulation = ______ inches thickness × ______ R-value per inch (table 15, text) = ______

R-value of duct = 0.8 (no insulation)
= R-value of insulation (if insulated)

Temperature difference between inside and outside of duct = ______ °F. temperature of hot air − ______ temperature °F. of attic or crawl space (table C-16 or 24, text) = ______ °F.

Heat loss = ______ ft^2 area × ______ °F. temperature difference × ______ hours of furnace on-time (table C-24) ÷ ______ R-value of duct = ______ Btu

Gas:

Ccf = ______ Btu heat loss ÷ ______ Btu per Ccf useful heat (table C-7) = ______ Ccf

Oil:

Gallons = ______ Btu heat loss ÷ ______ Btu per gallon useful heat (table C-7) = ______ gallon

Electricity:

Kwh = ______ Btu heat loss ÷ 3,413 Btu per kwh = ______ kwh

TABLE C-26
Total Energy Needed to Replace Heat Losses

	Heat Losses (in Ccf, gallon, or kwh)	*Cost of Heat Losses*
Windows (table C-11)	______	$______
Doors (table C-12)	______	______
Walls (table C-14)	______	______
Ceilings (table C-17)	______	______
Basement (table C-19)	______	______
or		
Crawl Space (table C-21)	______	______
or		
Slab Floor (table C-22)	______	______
Infiltration (table C-23)	______	______
Hot Air Ducts (table C-25)	______	______

Total Energy Needed to Replace Heat Losses = ______ {Ccf / gallon / kwh} $______

TABLE C-27
Expected Savings from Inside Temperature Reduction
("Rough Estimate")

Energy saved = ______ fractional[a] savings (table 27, text) × ______ {Ccf / gallon / kwh} used for heat (table C-8) = ______ {Ccf / gallon / kwh}

Savings per year = ______ {Ccf / gallon / kwh} saved × $______ per {Ccf / gallon / kwh} "last" rate (table C-6) = $______

[a] Divide the percent savings (text table 27) by 100 to convert to a fractional saving; that is, 24 percent becomes 0.24, 8 percent becomes 0.08, and so on.

TABLE C-28
Expected Savings from Inside Temperature Reduction ("Better value")

Heat Loss	*Before (Ccf, gallon, or kwh)*		*After (Ccf, gallon, or kwh)*		*Difference (Ccf, gallon, or kwh)*
Windows (table C-11)	______	–	______	=	______
Doors (table C-12)	______	–	______	=	______
Walls (table C-14)	______	–	______	=	______
Ceilings (table C-17)	______	–	______	=	______
Basement (table C-19)	______	–	______	=	______
or					
Crawl Space (table C-21)	______	–	______	=	______
or					
Slab Floor (table C-22)	______	–	______	=	______
Infiltration (table C-23)	______	–	______	=	______
Hot Air Ducts (table C-25)	______	–	______	=	______
			Total energy saved	=	______

TABLE C-29
Expected Savings from Adding Insulation ("Rough Estimate")

Ceiling:

Btu saved per year = ______ value from text table 28 × ______ area of ft² ceiling = ______ Btu

Gas:

Ccf saved = ______ Btu saved per year ÷ ______ Btu per Ccf useful heat (table C-7) = ______ Ccf

Oil:

Gallon saved = ______ Btu saved per year ÷ ______ Btu per gallon useful heat (table C-7) = ______ gallon

Electricity:

Kwh saved = ______ Btu saved per year ÷ 3,413 Btu per kwh = ______ kwh

Savings per year = ______ {Ccf / gallon / kwh} saved × $______ per {Ccf / gallon / kwh} "last" rate = $______ (table C-6)

Crawl Space:

Btu saved per year = ______ value from text table 28 × ______ ft^2 area of floor = ______ Btu

Gas:

Ccf saved = ______ Btu saved per year ÷ ______ Btu per Ccf useful heat (table C-7) = ______ gallon

Oil:

Gallon saved = ______ Btu saved per year ÷ ______ Btu per gallon useful heat (table C-7) = ______ Ccf

Electricity:

Kwh saved = ______ Btu saved per year ÷ 3,413 Btu per kwh = ______ kwh

Savings per year = ______ {Ccf / gallon / kwh} saved × $______ per {Ccf / gallon / kwh} "last" rate (table C-6) = $______

Walls:

Btu saved per year = ______ value from text table 28 × ______ ft^2 area of walls = ______ Btu

Gas:

Ccf saved = ______ Btu saved per year ÷ ______ Btu per Ccf useful heat (table C-7) = ______ gallon

Oil:

Gallon saved = ______ Btu saved per year ÷ ______ Btu per gallon useful heat (table C-7) = ______ gallon

Kwh saved = ______ Btu saved per year ÷ 3,413 Btu per kwh = ______ kwh

Savings per year = ______ {Ccf / gallon / kwh} saved × $______ per {Ccf / gallon / kwh} "last" rate (table C-6) = $______

TABLE C-30
Expected Savings from Adding Insulation ("Better Value")

Ceiling:

$$\text{Fuel} = \_\_\_\_ \left\{\begin{matrix}\text{Ccf}\\ \text{gallon}\\ \text{kwh}\end{matrix}\right\} \text{``Before''} \underset{\text{(table C-17 or C-24)}}{\text{heat loss}} - \_\_\_\_ \left\{\begin{matrix}\text{Ccf}\\ \text{gallon}\\ \text{kwh}\end{matrix}\right\} \text{``After'' heat loss} = \_\_\_\_ \left\{\begin{matrix}\text{Ccf}\\ \text{gallon}\\ \text{kwh}\end{matrix}\right\} \text{saved}$$

$$\text{Savings per year} = \_\_\_\_ \left\{\begin{matrix}\text{Ccf}\\ \text{gallon}\\ \text{kwh}\end{matrix}\right\} \text{saved} \times \$\_\_\_\_ \text{ per } \left\{\begin{matrix}\text{Ccf}\\ \text{gallon}\\ \text{kwh}\end{matrix}\right\} \underset{\text{(table C-6)}}{\text{``last'' rate}} = \$\_\_\_\_$$

Crawl Space:

$$\text{Fuel} = \_\_\_\_ \left\{\begin{matrix}\text{Ccf}\\ \text{gallon}\\ \text{kwh}\end{matrix}\right\} \text{``Before''} \underset{\text{(table C-21 or C-28)}}{\text{heat loss}} - \_\_\_\_ \left\{\begin{matrix}\text{Ccf}\\ \text{gallon}\\ \text{kwh}\end{matrix}\right\} \text{``After'' heat loss} = \_\_\_\_ \left\{\begin{matrix}\text{Ccf}\\ \text{gallon}\\ \text{kwh}\end{matrix}\right\} \text{saved}$$

$$\text{Savings per year} = \_\_\_\_ \left\{\begin{matrix}\text{Ccf}\\ \text{gallon}\\ \text{kwh}\end{matrix}\right\} \text{saved} \times \$\_\_\_\_ \text{ per } \left\{\begin{matrix}\text{Ccf}\\ \text{gallon}\\ \text{kwh}\end{matrix}\right\} \underset{\text{(table C-6)}}{\text{``last'' rate}} = \$\_\_\_\_$$

Walls:

$$\text{Fuel} = \_\_\_\_ \left\{\begin{matrix}\text{Ccf}\\ \text{gallon}\\ \text{kwh}\end{matrix}\right\} \text{``Before''} \underset{\text{(table C-14 or C-28)}}{\text{heat loss}} - \_\_\_\_ \left\{\begin{matrix}\text{Ccf}\\ \text{gallon}\\ \text{kwh}\end{matrix}\right\} \text{``After'' heat loss} = \_\_\_\_ \left\{\begin{matrix}\text{Ccf}\\ \text{gallon}\\ \text{kwh}\end{matrix}\right\} \text{saved}$$

$$\text{Savings per year} = \_\_\_\_ \left\{\begin{matrix}\text{Ccf}\\ \text{gallon}\\ \text{kwh}\end{matrix}\right\} \text{saved} \times \$\_\_\_\_ \text{ per } \left\{\begin{matrix}\text{Ccf}\\ \text{gallon}\\ \text{kwh}\end{matrix}\right\} \underset{\text{(table C-6)}}{\text{``last'' rate}} = \$\_\_\_\_$$

TABLE C-31
Expected Savings from Insulating Your Basement
("Rough Estimate")

Btu Saved

Band Joist:

Btu saved per year = ______ value from text table 28 × ______ ft length of band joist insulated = ______ Btu

Wall above ground:

Btu saved per year = ______ value from text table 28 × ______ ft² of insulation = ______ Btu

Wall below ground:

Btu saved per year = ______ value from text table 28 × ______ ft² of insulation = ______ Btu

Total saved per year = ______ Btu

Fuel and Money Saved

Gas:

Ccf saved = ______ Btu saved per year ÷ ______ Btu per Ccf useful heat (table C-7) = ______ Ccf

Oil:

Gallon saved = ______ Btu saved ÷ ______ Btu per gallon useful heat (table C-7) = ______ gallon

Electricity:

Kwh saved = ______ Btu saved per year ÷ 3,413 Btu per kwh = ______ kwh

Savings per year = ______ {Ccf / gallon / kwh} saved × $______ per {Ccf / gallon / kwh} "last" rate (table C-6) = $______

TABLE C-32
Expected Savings from Insulating Your Basement ("Better Value")

Fuel saved = ______ {Ccf / gallon / kwh} "Before" heat loss (table C-19 or C-28) − ______ {Ccf / gallon / kwh} "After" heat loss = ______ {Ccf / gallon / kwh}

Savings per year = ______ {Ccf / gallon / kwh} saved × $______ per {Ccf / gallon / kwh} "last" rate (table C-6) = $______

TABLE C-33
Fuel and Money Saved by Adding Insulation Around a Slab Floor ("Rough Estimate")

Btu saved per year = ______ value from text table 30 × ______ ft length of insulation added = ______ Btu

Fuel and Money Saved

Gas:

Ccf saved = ______ Btu saved per year ÷ ______ Btu per Ccf useful heat (table C-7) = ______ Ccf

Oil:

Gallon saved = ______ Btu saved per year ÷ ______ Btu per gallon useful heat (table C-7) = ______ gallon

Electricity:

Kwh saved = ______ Btu saved per year ÷ 3,413 = ______ Kwh

Savings per year = ______ {Ccf / gallon / kwh} saved × $______ per {Ccf / gallon / kwh} "last" rate (table C-6) = $______

TABLE C-34
Fuel and Money Saved by Adding Insulation Around a Slab Floor ("Better Value")

$$\text{Fuel saved} = \_\_\_\_\_ \begin{Bmatrix}\text{Ccf}\\ \text{gallon}\\ \text{kwh}\end{Bmatrix} \underset{\text{(table C-22 or C-28)}}{\text{``Before'' heat loss}} - \_\_\_\_\_ \begin{Bmatrix}\text{Ccf}\\ \text{gallon}\\ \text{kwh}\end{Bmatrix} \text{``After'' heat loss} = \_\_\_\_\_ \begin{Bmatrix}\text{Ccf}\\ \text{gallon}\\ \text{kwh}\end{Bmatrix}$$

$$\text{Savings} = \_\_\_\_\_ \begin{Bmatrix}\text{Ccf}\\ \text{gallon}\\ \text{kwh}\end{Bmatrix} \text{saved} \times \$\_\_\_\_\_ \text{ per } \begin{Bmatrix}\text{Ccf}\\ \text{gallon}\\ \text{kwh}\end{Bmatrix} \underset{\text{(table C-6)}}{\text{``last'' rate}} = \$\_\_\_\_\_$$

TABLE C-35
Expected Savings from Insulating Heat Ducts ("Better Value")

$$\text{Fuel saved} = \_\_\_\_\_ \begin{Bmatrix}\text{Ccf}\\ \text{gallon}\\ \text{kwh}\end{Bmatrix} \underset{\text{(table C-25)}}{\text{``Before'' heat loss}} - \_\_\_\_\_ \begin{Bmatrix}\text{Ccf}\\ \text{gallon}\\ \text{kwh}\end{Bmatrix} \text{``After'' heat loss} = \_\_\_\_\_ \begin{Bmatrix}\text{Ccf}\\ \text{gallon}\\ \text{kwh}\end{Bmatrix}$$

$$\text{Savings per year} = \_\_\_\_\_ \begin{Bmatrix}\text{Ccf}\\ \text{gallon}\\ \text{kwh}\end{Bmatrix} \text{saved} \times \$\_\_\_\_\_ \text{ per } \begin{Bmatrix}\text{Ccf}\\ \text{gallon}\\ \text{kwh}\end{Bmatrix} \underset{\text{(table C-6)}}{\text{``last'' rate}} = \$\_\_\_\_\_$$

TABLE C-36
Expected Savings from Adding Storm Windows and Doors ("Rough Estimate")

Windows:

Btu saved per year = _____ value from text table 31 × _____ ft^2 of added storm windows = _____ Btu

Doors:

Btu saved per year = _____ value from text table 31 × _____ ft^2 of added storm doors = _____ Btu

Total saved per year = _____ Btu

Fuel and Money Saved

Gas:

Ccf saved = ______ Btu saved per year ÷ ______ Btu per Ccf useful heat (table C-7) = ______ Ccf

Oil:

Gallon saved = ______ Btu saved per year ÷ ______ Btu per gallon useful heat (table C-7) = ______ gallon

Electricity:

Kwh saved = ______ Btu saved per year ÷ 3,413 Btu per kwh = ______ kwh

Savings per year = ______ {Ccf / gallon / kwh} saved × $______ per {Ccf / gallon / kwh} "last" rate (table C-6) = $______

TABLE C-37
Expected Savings from Adding Storm Windows and Doors ("Better Value")

Windows:

Fuel saved = ______ {Ccf / gallon / kwh} "Before" heat loss (table C-11 or C-28) − ______ {Ccf / gallon / kwh} "After" heat loss = ______ {Ccf / gallon / kwh}

Doors:

Fuel saved = ______ {Ccf / gallon / kwh} "Before" heat loss (table C-12 or C-28) − ______ {Ccf / gallon / kwh} "After" heat loss = ______ {Ccf / gallon / kwh}

Total fuel savings = ______ {Ccf / gallon / kwh}

Savings per year = ______ {Ccf / gallon / kwh} saved × $______ per {Ccf / gallon / kwh} "last" rate (table C-6) = $______

TABLE C-38
Expected Savings from Caulking and Weather Stripping

Cubic feet per hour reduction = ______ ft length of weather stripping[a] × ______ leakage savings factor (text table 32) = ______ cfh

Cubic feet per hour reduction[b] = ______ ft length of weather stripping × ______ leakage savings factor = ______ cfh

Cubic feet per hour reduction[b] = ______ ft length of weather stripping × ______ leakage savings factor = ______ cfh

Total infiltration reduction = ______ cfh

Btu saved = ______ cfh reduction × ______ Ddh (table C-10)[c] × 0.018 = ______ Btu

Fuel and Money Saved

Gas:

Ccf saved = ______ Btu saved per year ÷ ______ Btu per Ccf useful heat (table C-7) = ______ Ccf

Oil:

Gallon saved = ______ Btu saved per year ÷ ______ Btu per gallon useful heat (table C-7) = ______ gallon

Electricity:

Kwh saved = ______ Btu saved per year ÷ 3,413 Btu per kwh = ______ kwh

Savings per year = ______ {Ccf / gallon / kwh} saved × $______ per {Ccf / gallon / kwh} "last" rate (table C-6) = $______

---

[a] If you add a storm door, the length is the perimeter of the door. If you add a storm window to a double-hung sash, the length is twice the height plus three times the width.

[b] Use as many factors as necessary.

[c] If you also plan to reduce the inside temperature, you must recalculate the degree-difference-hours at the new inside temperature (table C-10).

TABLE C-39
Btu's Required to Heat a Gallon of Water

Temperature rise	= ______	°F. of hot-water temperature	− ______	°F. of cold-water temperature	= ______	°F.	
Btu added to 1 gallon hot water	= ______	°F. temperature rise	×	8.34 Btu per gallon per °F.	= ______	Btu	

TABLE C-40
Cost of Hot Water from Electric Heater

Kwh required to heat 1 gallon	= ______	Btu in 1 gallon (table C-39)	÷ 3,413 Btu per kwh	= ______	kwh	
Cost per gallon	= ______	kwh added to 1 gallon hot water	× $______	per kwh average rate (table C-6)	= $______	per gallon
Money saved per gallon saved	= ______	kwh added to 1 gallon hot water	× $______	per kwh "last" rate (table C-6)	= $______	per gallon

TABLE C-41
Cost of Hot Water from Oil or Gas Heater

Btu per hour for hot water	= 8.34 Btu per gallon per °F.	× ______	gallon per hour recovery rate			
		× ______	°F. temperature rise	= ______	Btu per hour	
Efficiency	= ______	Btu per hour for hot water	÷ ______	Btu-per-hour input[a]	= 0.______	(=______%)[b]
Btu to heat 1 gallon water	= ______	Btu added to 1 gallon hot water (table C-39)	÷ 0.______	efficiency	= ______	Btu

Gas:

Ccf to heat 1 gallon water	= ______	Btu to heat 1 gallon water	÷ 100,000 Btu per Ccf	= ______ Ccf

Cost per gallon of water	= ______	Ccf of gas to heat 1 gallon water	× $______ per Ccf average rate (table C-6)	= $______ per gallon
Savings per gallon of hot water saved	= ______	Ccf of gas to heat 1 gallon water	× $______ per Ccf "last" rate (table C-6)	= $______ per gallon

Oil:

gallons to heat 1 gallon water	= ______	Btu to heat 1 gallon water	÷ 140,000 Btu per gallon	= ______ gallon

Cost per gallon of water	= ______	gallon oil to heat 1 gallon water	× $______ per gallon oil (table C-6)	= $______ per gallon

Savings per gallon saved = cost per gallon water (same price)

---

[a] If your gas hot-water heater has no Btu per hour input listed, calculate that value. Use the test dial method (see chapter 2) and do the calculation in the space below.

Time for one revolution of 1/2-cubic-foot dial	= ______ seconds ÷ 60	= ______ minutes		
Gas consumption (including all pilot lights)	= 30 ÷ ______	minute for one revolution of 1/2-cubic-foot dial	= ______ cfh	
Gas consumption of heater only	= ______	cfh gas consumption (including pilot lights)	− ______ cfh for pilot lights (table C-2)	= ______ cfh

[b] To find the percent efficiency, multiply the previous fraction by 100.

TABLE C-42
Hot-Water Flow Rates for Bathtub(s)

Hot-water flow Tub #1	= {60 (gallon container) / 15 (quart container)}	÷ ______	seconds to fill	= ______ gpm
Hot-water flow Tub #2	= {60 (gallon container) / 15 (quart container)}	÷ ______	seconds to fill	= ______ gpm

TABLE C-43
Hot-Water Flow Rates for Shower(s)

Temperature of cold water = ______ °F. Temperature of hot water = ______ °F.

Person: ______________________________

Warm-water flow = {60 (gallon container) / 15 (quart container)} ÷ ______ seconds to fill = ______ gpm

Temperature of warm water = ______ °F.

Temperature difference between warm and cold water = ______ °F. warm water − ______ °F. cold water = ______ °F.

Temperature difference between hot and cold water = ______ °F. hot water − ______ °F. cold water = ______ °F.

Hot-water fraction = ______ °F. temperature difference between warm and cold water ÷ ______ °F. temperature difference between hot and cold water = 0.______

Hot-water flow = ______ hot-water fraction × ______ gpm warm-water flow = ______ gpm

[a]Person: ____________________: Hot-water flow = ______ gpm
Person: ____________________: Hot-water flow = ______ gpm
Person: ____________________: Hot-water flow = ______ gpm
Person: ____________________: Hot-water flow = ______ gpm

[a] If different family members have different hot-water flow rates, calculate the flow rate for each person using the method of calculation shown.

TABLE C-44
Hot Water Used for Showers and Baths by Your Family

Person				
______	Shower hot water = ______	gpm hot water × ______ (table C-43)	minutes = ______	gallons
	Bath hot water = ______	gpm hot water × ______ (table C-42)	minutes = ______	gallons
______	Shower hot water = ______	gpm hot water × ______	minutes = ______	gallons
	Bath hot water = ______	gpm hot water × ______	minutes = ______	gallons
______	Shower hot water = ______	gpm hot water × ______	minutes = ______	gallons
	Bath hot water = ______	gpm hot water × ______	minutes = ______	gallons
______	Shower hot water = ______	gpm hot water × ______	minutes = ______	gallons
	Bath hot water = ______	gpm hot water × ______	minutes = ______	gallons
______	Shower hot water = ______	gpm hot water × ______	minutes = ______	gallons
	Bath hot water = ______	gpm hot water × ______	minutes = ______	gallons

TABLE C-45
Total Hot Water Used by Your Family

	Gallons per day[a]	*Cost per day ($)*[b]	*Gallons per month*[c]	*Cost per month ($)*[c]
Showers and baths (table C-44)				
Person: ________	________	________	________	________
Person: ________	________	________	________	________
Person: ________	________	________	________	________
Person: ________	________	________	________	________
Person: ________	________	________	________	________
Handwashings (Times per day × gallons per use)				
Person: ________	________	________	________	________
Person: ________	________	________	________	________
Person: ________	________	________	________	________
Person: ________	________	________	________	________
Person: ________	________	________	________	________
Dishwashing[d]				
By hand	________	________	________	________
Dishwasher	________	________	________	________
Clothes washer[e]	________	________	________	________
Other uses				
________	________	________	________	________
________	________	________	________	________
________	________	________	________	________
________	________	________	________	________
________	________	________	________	________
Totals	________ gallons	$ ________	________ gallons	$ ________

[a] If any use is more or less frequent than once a day, calculate the use on a per-day basis.

[b] Multiply the gallons of hot water per day by the average cost per gallon of hot water (table C-41).

[c] Multiply gallons per day and cost per day by 30 to determine the gallons and cost per month.

[d] See later section on dishwashers (table C-80 or C-84).

[e] See later section on clothes washers (table C-100).

TABLE C-46
Calculation of Standby Heat Loss

R-value = 3 for 1 inch of insulation
= 5 for 2 inches of insulation

Temperature difference	=	______	°F. hot-water temperature	−	______	°F. room temperature (where heater is located)	= ______	°F.

Btu heat loss per month	=	______	ft² area (text table 34)	×	______	°F. temperature difference	÷	______	R-value
							=	______	Btu per hour

Btu heat loss per month	=	______	Btu heat loss per hour	×	732 hours per month	=	______	Btu

Energy Required to Replace Standby Heat Loss

Electric:

Kwh to replace standby heat loss per month	=	______	Btu heat loss per month	÷ 3,413 Btu per kwh	=	______	kwh

Gas:

Btu per Ccf useful heat	=	100,000 Btu per Ccf	×	0.______	efficiency (table C-41)	= ______	Btu per Ccf

Ccf to replace standby heat loss per month	=	______	Btu heat loss per month	÷	______	Btu per Ccf useful heat	= ______	Ccf

Oil:

Btu per gallon useful heat	=	140,000 Btu per gallon	×	0.______	efficiency (table C-41)	= ______	Btu per gallon

Gallon to replace standby heat loss per month	=	______	Btu heat loss per month	÷	______	Btu per gallon useful heat	= ______	gallon

Cost per month	=	______	{Ccf / gallon / kwh}	to replace standby heat loss per month	× $______	per	{Ccf / gallon / kwh}	average rate (table C-6) = $______

TABLE C-47
Standby Heat Loss Calculation: On-time–Off-time Method

On-time = ______ minutes ÷ 60 minutes per hour = ______ hour

Off-time = ______ hours ______ minutes = ______ hours[a]

Cycle time ______ hour on-time + ______ hours off-time = ______ hour

Electric heater: Input = ______ watts[b] ÷ 1000 = ______ kilowatts

Energy used each on-time = ______ hours on-time × ______ {kilowatts / Btu per hour / input[c]} = ______ {kwh / Btu}

Energy to replace standby heat = ______ {kwh / Btu} used each on-time ÷ ______ hours cycle-time

= ______ {kwh / Btu} per hour

Energy to replace standby heat loss per month = ______ {kwh / Btu} to replace standby heat loss per hour × 732 hours per month = ______ {kwh / Btu}

Gas:

Ccf to replace standby heat loss = ______ Btu to replace standby heat loss per month ÷ 100,000 Btu per Ccf = ______ Ccf

Oil:

Gallons to replace standby heat loss per month = ______ Btu to replace standby heat loss per month ÷ 140,000 Btu per gallon

= ______ gallons

Cost per month = ______ {Ccf / gallon / kwh} to replace standby heat loss per month × $______ per {Ccf / gallon / kwh} average rate (table C-6)

= $______

---

[a] Divide the minutes by 60 to determine the fraction of an hour.

[b] If the wattage is not listed or if your heater has two heating elements and neither wattage is listed, calculate the wattage of the element that turns on. Use the rotating disc method (see chapter 2), then do the calculation in the space below.

Time required for one disc revolution = ______ seconds

Watts = 3,600 × Kh value (table C-1) ÷ ______ seconds for one disc revolution = ______ watts

[c] Table C-41.

TABLE C-48
Standby Heat Loss: Vacation Method

Electric:

Kwh used (heater on) = ______ kwh final meter reading − ______ kwh initial meter reading = ______ kwh

Kwh per hour (heater on) = ______ kwh used (heater on) ÷ ______ hours gone = ______ kwh per hour

Kwh used (heater off) = ______ kwh final meter reading − ______ kwh initial meter reading = ______ kwh

Kwh per hour (heater off) = ______ kwh used (heater off) ÷ ______ hours gone = ______ kwh per hour

Standby heat loss per hour = ______ kwh per hour (heater on) − ______ kwh per hour (heater off) = ______ kwh per hour

Standby heat loss per month = ______ kwh standby heat loss per hour × 732 hours per month = ______ kwh

Cost per month = ______ kwh standby heat loss per month × $______ per kwh average rate (table C-6) = $______

Gas:

Ccf used = ______ Ccf final meter reading − ______ Ccf initial meter reading = ______ Ccf

Ccf total per hour = ______ Ccf used ÷ ______ hours gone = ______ Ccf per hour

Standby heat loss per hour = ______ Ccf total per hour − ______ Ccf per hour pilot lights (table C-2) = ______ Ccf per hour

Standby heat loss per month = ______ Ccf standby heat loss per hour × 732 hours per month = ______ Ccf per month

Cost per month = ______ Ccf per month × $______ per Ccf average rate (table C-6) = $______

TABLE C-49
Water Needed to Flush Cooled Water from Hot-Water Pipes

	Uses per Day	*Gallons[a] to Flush Pipe*	*Gallons per Day[b]*	*Gallons per Month[c]*
Bathroom #1				
Sink	______	______	______	______
Shower/tub	______	______	______	______
Bathroom #2				
Sink	______	______	______	______
Shower/tub	______	______	______	______
Bathroom #3				
Sink	______	______	______	______
Laundry[d]	______	______	______	______
Kitchen[d]	______	______	______	______
		Totals	______ gallons	______ gallons

Cost of pipe losses per month = ______ gallons lost per month × $______ per gallon average cost (table C-40 or C-41) = $______

[a] If you use a cup measure, convert cups to gallons by dividing the number of cups of water by 16; if you use a pint measure, divide number of pints by 8 to determine number of gallons.

[b] Gallons per day = uses per day × gallons to flush pipe.

[c] Gallons per month = gallons per day × 30 days per month.

[d] Include only the laundry and kitchen sink uses; do not include the clothes washer and dishwasher. All the water in the hot-water pipe enters the dishwasher and clothes washer—none is flushed out before they are turned on. The inclusion of the cooled water in the pipe results in a cooler water temperature in the washer; no additional hot water is used.

TABLE C-50
Hot Water Saved by Reducing Shower and Bath Water Consumption

"After" baths and showers:

Person	*Gpm Hot-water Flow*		*Minutes Hot Water Used*		*Number of Showers or Baths per Month*		*Gallons Used per Month*
______:	______	×	______	×	______	=	______
______:	______	×	______	×	______	=	______
______:	______	×	______	×	______	=	______
______:	______	×	______	×	______	=	______
					Total	=	______ gallons

Hot water saved	= ______	gallons used per month "before" (table C-45)	− ______	gallons used "after"	= ______	gallons saved per month

Savings per month	= ______	gallons saved per month	× $______	per gallon hot water savings rate (table C-40 or C-41)	= $______

TABLE C-51
Hot Water Saved by Reducing Dish Rinsing: Dishwasher

Gallons of hot water saved per day	= ______	gallons used for rinsing per day	− ______	gallons used for no rinsing or very little rinsing
				= ______ gallons

Gallons saved per month	= ______	gallons saved per day	× 30 = ______	gallons

Savings per month	= ______	gallons saved per month	× $______	per gallon hot water savings rate (tables C-40 or C-41) = $______

TABLE C-52
Hot Water Saved by Reducing Dish Rinsing: Hand Washing

Gallons of hot water saved per day	= ______	gallons used per day "before"	− ______	gallons used per day "after"	= ______ gallons

Gallons saved per month = ______ gallons saved per day × 30 = ______ gallons

Savings per month	= ______	gallons saved per month	× $______	per gallon hot water savings rate (tables C-40 or C-41) = $______

TABLE C-53
Hot Water Saved by Repairing Leaky Faucet

Gallons saved per month by repair	= ______	cups per hour from dripping faucet	×	732 hours per month	÷ 16 cups per gallon
				= ______	gallons per month

Savings per month	= ______	gallons saved per month	× $______	per gallon hot water savings rate (tables C-40 or C-41)	= $______

TABLE C-54
Hot Water Saved by Miscellaneous Reductions

Use: ____________ gallons saved per month = ______ gallons
Use: ____________ gallons saved per month = ______ gallons
Use: ____________ gallons saved per month = ______ gallons

Total = ______ gallons

Savings per month	= ______	gallons saved per month	× $______	per gallon hot water savings rate (tables C-40 or C-41)	= $______

TABLE C-55
Expected Savings from a Reduction of Hot-Water Temperature

Temperature reduction	= ______	°F. previous hot-water temperature	− ______	°F. new hot-water temperature	= ______ °F.

Fractional savings = ______ °F. temperature reduction × 0.003 = 0.______

Savings = $______ for hot water[a] (table C-45) × 0.______ fractional savings = $______

Additional energy savings from standby heat loss:

Previous temperature difference between hot water and room	= ______	°F. previous hot-water temperature	− ______	°F. room temperature	= ______ °F.		

Fractional savings (standby heat loss)	= ______	°F. temperature reduction	÷ ______	°F. previous temperature difference between hot water and room	= 0.______

Savings per month	= 0.______	fractional savings (standby heat loss)	× $______	monthly cost of standby heat loss (tables C-46, C-47, C-48)	= $______

[a] If you follow any of the previous conservation measures, the cost for your hot water is the original cost (table C-45) minus the savings from any of the conservation measures that you use.

TABLE C-56
Expected Savings from Turning Off Hot Water During Vacations

Standby heat loss saved	= $______	standby heat loss cost per month (tables C-46, C-47 or C-48)	× ______	days gone	÷ 30 days per month	= $______

Cost to reheat water in tank	= ______	gallon capacity of tank	× $______	per gallon average rate (table C-40 or C-41)	= $______

Actual savings	= $______	standby heat loss saved	− $______	to reheat water	= $______

TABLE C-57
Monthly Energy Use of Refrigerator and Freezer
(*AHAM Directory*)

Refrigerator energy use per month = ______ kwh
Freezer energy use per month = ______ kwh

TABLE C-58
Monthly Energy Use of Refrigerator and Freezer: On-time–Off-time Method

Refrigerator:

Wattage (use rotating disc method):

Time for one disc revolution = ______ seconds

Wattage = 3,600 × ______ Kh value (table C-1) ÷ ______ seconds for one disc revolution = ______ watts

Times:

Time on:[a] ______ Time off:[a] ______ Time on again:[a] ______

On-time = ______ time off − ______ time on = ______ minutes

Cycle time = ______ time on again − ______ time on = ______ minutes

Fraction of time on = ______ minutes on-time ÷ ______ minutes cycle time = 0.______

Heat gain per month = ______ watts × 0.______ fraction of time on × 732 ÷ 1000 = ______ kwh

Energy used per month = ______ kwh heat gain × 1.2 = ______ kwh

Freezer:

Wattage (use rotating disc method):

Time for one disc revolution = ______ seconds

Wattage = 3,600 × ______ Kh value (table C-1) ______ seconds for one disc revolution = ______ watts

Times:

Time on:[a] ______ Time off:[a] ______ Time on again:[a] ______

On-time = ______ time off − ______ time on = ______ minutes

Cycle time = ______ time on again − ______ time on = ______ minutes

Fraction of time on = ______ minutes on-time ÷ ______ minutes cycle time = 0.______

Heat gain per month = ______ watts × 0.______ fraction of time on × 732 ÷ 1000 = ______ kwh

Energy used per month = ______ kwh heat gain per month × 1.1 = ______ kwh

[a] Record actual clock time here.

TABLE C-59
Monthly Energy Use of Refrigerator and Freezer: Electric Meter Method

Energy required to remove heat gain from conduction:

Kwh used during selected period = ______ kwh final meter reading − ______ kwh initial meter reading = ______ kwh

Heat gain by refrigerator (and freezer) during period = ______ kwh used during period − ______ kwh used by other appliances[a] = ______ kwh

Kwh used by refrigerator (and freezer) per month = ______ kwh heat gain by refrigerator (and freezer) during period × 732 ÷ ______ hours of selected period = ______ kwh per month

Include heat gain from warm food and warm air entering through open door:

If you have *only* a refrigerator:

Total energy used per month = ______ kwh heat gain per month × 1.2 = ______ kwh

If both refrigerator and freezer are included and the energy used by each can't be found separately:

Total energy used by refrigerator and freezer = ______ kwh heat gain by refrigerator and freezer × 1.15 = ______ kwh

If one is a manual-defrost and you use on-time—off-time method to calculate the heat gain:

Refrigerator (freezer is manual-defrost):

Heat gain of refrigerator = ______ kwh used by refrigerator and freezer per month − ______ kwh freezer heat gain per month (table C-57) = ______ kwh

Energy used per month = ______ heat gain of refrigerator per month × 1.2 = ______ kwh per month

Freezer (refrigerator is manual-defrost):

Heat gain of freezer per month = ______ kwh used by refrigerator and freezer per month − ______ kwh refrigerator heat gain per month (table C-57) = ______ kwh

Energy used per month = ______ kwh heat gain per month × 1.1 = ______ kwh per month

[a] For appliances such as lights, multiply the wattage by the on-time (in hours) and divide by 1,000 to obtain the kilowatt-hour consumption.

TABLE C-60
Expected Savings by Moving Refrigerator or Freezer to a Cooler Room

Temperature change = ______ °F. present room temperature − ______ °F. new room temperature = ______ °F.[a]

Fractional decrease of energy = ______ °F. temperature change × {0.02 (refrigerator) / 0.015 (freezer)} = 0.______

Energy savings per month = ______ kwh used per month (table C-57, C-58 or C-59) × 0.______ fractional decrease of energy = ______ kwh

Repeat this calculation for the different months in which room temperature is different. Record the savings in the following spaces.

Month	*Kwh*[b] *Saved*	*Month*	*Kwh*[b] *Saved*	*Month*	*Kwh*[b] *Saved*
______	______	______	______	______	______
______	______	______	______	______	______
______	______	______	______	______	______
______	______	______	______	______	______

Total energy saved = ______ kwh

Savings = ______ kwh saved × $______ per kwh "last" rate (table C-6) = $______

[a] If this value is negative (that is, new temperature is higher than the old temperature), this is an *increase* of energy use.

[b] If any of these values are negative (that is, an energy increase rather than an energy savings), subtract those values rather than add them to find the total.

TABLE C-61
Expected Savings from Purchase of a More Efficient Refrigerator or Freezer

Energy saved per month = ______ kwh per month of present unit (table C-57, C-58 or C-59) − ______ kwh per month of new unit = ______ kwh

Savings per month = ______ kwh saved per month × $______ per kwh "last" rate (table C-6) = $______

TABLE C-62
Wattages for Bake and Broil Elements: Rotating Disc Method

Preheat (broil and bake elements together):

Time for one disc revolution = ______ seconds

Wattage = 3,600 × ______ Kh value (table C-1) ÷ ______ seconds for one disc revolution = ______ watts

Bake Element Only:

Time for one disc revolution = ______ seconds

Wattage = 3,600 × ______ Kh value (table C-1) ÷ ______ seconds for one disc revolution = ______ watts

Broil Element Only:[a]

Time for one disc revolution = ______ seconds

Wattage = 3,600 × ______ Kh value (table C-1) ÷ ______ seconds for one disc revolution = ______ watts

[a] Set oven to broil and measure the time required for one disc revolution.

TABLE C-63
On-Times of Your Oven Heating Elements

To Preheat:

	Time[a] *hr:min:sec*	*Time on*[b] *min:sec*	*Time on*[c] *minutes*
Oven on:	__:___:__		
Broil off:	__:___:__	___:__	__.___ (broil only)
Bake off:	__:___:__	___:__	__.___ (bake only)

To Maintain 350° F.:

Estimate fraction of bake element on-time to maintain 350° F. = 0.____

or

On-time of bake element during one cycle = _____ minutes, _____ seconds = __.___ minutes on[c]

Time off during one cycle = _____ minutes, _____ seconds = __.___ minutes off[c]

Total time for one cycle = _____ minutes on + _____ minutes off

= _____ minutes

Fraction of on-time = _____ minutes on during one cycle ÷ _____ minutes total time for one cycle = 0.____

[a] Record actual clock times in this column.

[b] Subtract oven on-time from the time each element turns off.

[c] Convert the times from minutes and seconds to minutes and fractions of minutes by dividing the number of seconds by 60. For example, 5 minutes and 25 seconds is 5 minutes + 25/60 minutes = 5.42 minutes.

TABLE C-64
Energy Used by Electric Oven

Preheat:

Broil element energy	= ______	minutes broiler on (table C-63)	× ______	watts of broil element (table C-62)	÷ 60,000 = ______ kwh
Bake element energy	= ______	minutes bake element on (table C-63)	× ______	watts of bake element (table C-62)	÷ 60,000 = ______ kwh

Total preheat energy = ______ kwh

Power to maintain 350° F.	= ______	watts of bake element (table C-62)	× 0.______	fraction of time on (table C-63)	÷ 1000 = ______ kw

TABLE C-65
Energy Consumption of Gas Oven

Time required for one revolution of 1/2-cubic-foot dial	= ______	seconds ÷ 60 = ______ minutes	
Gas consumption (including all pilot lights)	= 30 ÷ ______	minutes for one revolution of 1/2-cubic-foot dial = ______ cfh	
Gas consumption of oven only	= ______	cfh gas consumption (including all pilot lights)	− ______ cfh for pilot lights (table C-2) = ______ cfh

TABLE C-66
On-Times of Oven Burner

On-time of burner during preheat cycle = ______ minutes ______ seconds = ___.___ minutes[a]

= ______ hours[b]

Estimated fraction of burner on-time required to maintain 350° F. = 0.______

OR

Burner on-time during one cycle = ______ minutes ______ seconds = ___.___ minutes[a] on

Burner off-time during one cycle = ______ minutes ______ seconds = ___.___ minutes[a] off

Total time for one cycle = ______ minutes on + ______ minutes off = ______ minutes

Fraction of on-time = ______ minutes on during one cycle ÷ ______ minutes total time for one cycle = 0.______

[a] Divide the number of seconds by 60 to convert to fractions of minutes. For example, 5 minutes and 25 seconds is 5 minutes plus 25/60 minutes = 5.42 minutes.

[b] Divide minutes by 60 to convert to hours.

TABLE C-67
Energy Used by Gas Oven

Preheat energy = ______ hours to preheat (table C-66) × ______ cfh gas consumption of oven only (table C-65) = ______ cf

Power to maintain 350° F. = ______ cfh gas consumption of oven only (table C-65) × 0.______ fraction of time on (table C-66) = ______ cfh

TABLE C-68
Energy Required to Preheat and to Maintain Different Oven Temperatures

Preheat Energy:

Temperature	*Energy Needed to Preheat Oven to 350° F. (table C-64 or C-67) cf or kwh*		*Energy Needed to Preheat Oven to Desired Temperature cf or kwh*
250° F.:	________	× 0.65 =	________
275° F.:	________	× 0.75 =	________
300° F.:	________	× 0.85 =	________
325° F.:	________	× 0.90 =	________
350° F.:	________	× 1.00 =	________
375° F.:	________	× 1.10 =	________
400° F.:	________	× 1.20 =	________
425° F.:	________	× 1.25 =	________
450° F.:	________	× 1.35 =	________

Desired Temperature Maintenance Power:

Temperature	*Power Needed to Maintain 350° F. (table C-64 or C-67) cfh or kw*		*Power Needed to Maintain Desired Temperature cfh or kw*
250° F.:	________	× 0.65 =	________
275° F.:	________	× 0.75 =	________
300° F.:	________	× 0.85 =	________
325° F.:	________	× 0.90 =	________
350° F.:	________	× 1.00 =	________
375° F.:	________	× 1.10 =	________
400° F.:	________	× 1.20 =	________
425° F.:	________	× 1.25 =	________
450° F.:	________	× 1.35 =	________

TABLE C-69
Energy Required to Prepare Various Foods in Oven

Food: ______________________ Temperature: __________ °F.

Energy needed to maintain temperature = ______ hours in oven × ______ $\begin{Bmatrix} \text{cfh} \\ \text{kw} \end{Bmatrix}$ (table C-68) = ______ $\begin{Bmatrix} \text{cf} \\ \text{kwh} \end{Bmatrix}$

Total energy = ______ $\begin{Bmatrix} \text{cf} \\ \text{kwh} \end{Bmatrix}$ to maintain temperature + ______ $\begin{Bmatrix} \text{cf} \\ \text{kwh} \end{Bmatrix}$ preheat energy (table C-68) = ______ $\begin{Bmatrix} \text{cf} \\ \text{kwh} \end{Bmatrix}$

Repeat the calculation as shown to find the total amount of energy needed to prepare other foods.

Food	*Temperature (°F.)*	*Total Energy (cf or kwh)*

TABLE C-70
Power of a Microwave Oven

Power rating = ______ watts ÷ 1,000 = ______ kilowatts (kw)
(from rear
of oven)

Variable cooking speeds (use rotating disc method):

Setting	*Effective Wattage*[a]		*Kilowatts (kw)*
______	______	÷ 1,000 =	______
______	______	÷ 1,000 =	______
______	______	÷ 1,000 =	______
______	______	÷ 1,000 =	______
______	______	÷ 1,000 =	______
______	______	÷ 1,000 =	______

[a] Repeat the calculations given below for each setting:

Time for one disc revolution = ______ seconds

Wattage = 3,600 × ______ Kh value (table C-1) ÷ ______ seconds for one disc revolution = ______ watts

TABLE C-71
Maximum Heat Outputs of Cook-Top Elements and Burners

Electric:

Large (use rotating disc method):

Time for one disc revolution = ______ seconds

Wattage = 3,600 × ______ Kh value (table C-1) ÷ ______ seconds for one disc revolution = ______ watts

Kilowatts = ______ watts ÷ 1,000 = ______ kw

Small (use rotating disc method):

Time for one disc revolution = ______ seconds

Wattage = 3,600 × ______ Kh value (table C-1) ÷ ______ seconds for one disc revolution = ______ watts

Kilowatts = ______ watts ÷ 1,000 = ______ kw

Gas:

If gas consumption is listed as Btu per hour, convert to cubic feet per hour (cfh):

High heat:
Maximum output = ______ Btu per hour ÷ 1,000 = ______ cfh

Medium heat:
Maximum output = ______ Btu per hour ÷ 1,000 = ______ cfh

If not listed, find the cubic feet per hour:

High heat:

Maximum output of burner plus gas for all pilot lights = 30 ÷ ______ minutes for one revolution of half-cubic-foot dial = ______ cfh

Maximum output of burner alone = ______ cfh (burner plus pilot lights) − ______ cfh of all pilot lights (table C-2)

= ______ cfh

Medium heat:

Maximum output of burner plus gas for all pilot lights = 30 ÷ ______ minutes for one revolution of half-cubic-foot dial = ______ cfh

Maximum output of burner alone = ______ cfh (burner plus pilot lights) − ______ cfh of all pilot lights (table C-2)

= ______ cfh

TABLE C-72
Energy Produced by Cook-Top Units at Different Settings

Setting	*Fraction of Maximum (text table 38)*		*Maximum Output (kw or cfh) (table C-71)*		*Output at Setting (kw or cfh)*
High heat or large heating element:					
______	___.______	×	________	=	________
______	___.______	×	________	=	________
______	___.______	×	________	=	________
______	___.______	×	________	=	________
______	___.______	×	________	=	________
______	___.______	×	________	=	________
______	___.______	×	________	=	________

Setting	*Fraction of Maximum (text table 38)*		*Maximum Ouput (kw or cfh) (table C-71)*		*Output at Setting (kw or cfh)*
Medium heat or small heating element:					
______	.______	×	________	=	________
______	.______	×	________	=	________
______	.______	×	________	=	________
______	.______	×	________	=	________
______	.______	×	________	=	________
______	.______	×	________	=	________
______	.______	×	________	=	________

TABLE C-73
Energies Required to Cook Some Foods on Cook Top

Food: ______________________________

Initial setting: ____________

Energy = ______ {kw / cfh} at this setting (table C-72) × ______ hours to cook = ______ {kwh / cf}

Final setting: ____________

Energy = ______ {kw / cfh} at this setting (table C-72) × ______ hours to cook = ______ {kwh / cf}

Total energy = ______ {kwh / cf}

Food	*Total Energy*[a]
______________________________	____________
______________________________	____________
______________________________	____________
______________________________	____________
______________________________	____________
______________________________	____________
______________________________	____________
______________________________	____________
______________________________	____________

[a] Repeat the procedure shown for other foods prepared on the stove and record the total energy required for each.

TABLE C-74
Energy Required to Operate Various Kitchen Appliances with Thermostats

Appliance: ________________ Temperature:[a] ____________

Preheat energy = ______ watts × ______ hours to preheat to desired temperature ÷ 1,000 = ______ kwh

Power needed to maintain desired temperature – ______ watts × 0.______ fraction of electricity on-time needed to maintain temperature[b] ÷ 1,000 = ______ kw

Appliance	*Temperature*[a]	*Preheat Energy (kwh)*[c]	*Power Needed to Maintain Temperature (kw)*[c]
________	________	________	________
________	________	________	________
________	________	________	________
________	________	________	________
________	________	________	________
________	________	________	________
________	________	________	________
________	________	________	________
________	________	________	________
________	________	________	________

[a] Temperature can be expressed either in terms of degrees F. or in terms of a setting number or word.

[b] Either estimate the fraction of electricity on-time or use the method outlined in table C-63 to calculate the fraction.

[c] Repeat the calculations shown for different appliances at different temperatures.

TABLE C-75
Energy Required to Prepare Foods on Other Kitchen Appliances

Appliances with thermostats:

Food: ________________ Appliance: ________________ Temperature: ______

Energy required to cook food = ______ hours needed to cook × ______ kw power needed to maintain temperature (table C-74) = ______ kwh

Preheat energy (table C-74) = ______ kwh

Total energy = ______ kwh

Appliances without thermostats:

Food: ______ Appliance: ______ Temperature: ______

Energy required to cook food = ______ hours needed to cook × ______ watts ÷ 1,000 = ______ kwh

Food	*Energy (kwh)*[a]	*Food*	*Energy (kwh)*
______	______	______	______
______	______	______	______
______	______	______	______
______	______	______	______
______	______	______	______
______	______	______	______
______	______	______	______
______	______	______	______
______	______	______	______

[a] Do one of the calculations shown for each food prepared on a kitchen appliance.

TABLE C-76
Total Monthly Energy Used in the Kitchen

Food[a]	*Energy per Preparation (Kwh or Cf) (table C-69, C-73, or C-75)*		*Preparations per Month*		*Kwh or Cf per Month*
______	______	×	______	=	______
______	______	×	______	=	______
______	______	×	______	=	______
______	______	×	______	=	______
______	______	×	______	=	______
______	______	×	______	=	______
______	______	×	______	=	______
______	______	×	______	=	______
______	______	×	______	=	______
______	______	×	______	=	______
______	______	×	______	=	______
______	______	×	______	=	______
______	______	×	______	=	______
______	______	×	______	=	______
______	______	×	______	=	______
______	______	×	______	=	______
______	______	×	______	=	______
______	______	×	______	=	______
______	______	×	______	=	______
______	______	×	______	=	______
			Total	=	______ {kwh / cf}

If you have a gas oven, cook-top, or both with pilot lights, add the pilot-light energy:

Pilot-light consumption per month = ______ number of pilot lights in oven, cook top, or both × ______ cf per month consumed per pilot light (table C-2) = ______ cf

Total gas for cooking = ______ cf per month to prepare food + ______ cf per month for pilot light(s) = ______ cf

[a] For foods such as cakes and pies or vegetables, don't list each different kind prepared; estimate the average baking or cooking time for all the different kinds and use that figure plus average temperature or cook-top setting to find the average energy.

TABLE C-77
Expected Savings from Using a Microwave Oven

Food: ______________________________

Energy used by microwave oven[a] = ______ watts of microwave oven × ______ hours to cook in microwave oven ÷ 1,000

= ______ kwh

vs. Electric Oven:

Kwh saved = ______ kwh used by electric oven (table C-69) − ______ kw used by microwave oven = ______ kwh

Savings = ______ kwh saved × $______ per kwh "last" rate (table C-6) = $______

vs. Gas Oven:

Cost to use gas oven = ______ cf used by gas oven (table C-69) × $______ per Ccf "last" rate (table C-6) ÷ 100 = $______

Cost to use microwave oven = ______ kwh used by microwave oven × $______ per kwh "last" rate (table C-6) = $______

Savings = $______ to use gas oven − $______ to use microwave oven = $______

Repeat this calculation for all the foods you could prepare in a microwave oven and record those savings in the following chart.

Food	*$ Saved per Preparation*		*Preparations per Month*		*Total Savings*
______	______	×	______	=	$______
______	______	×	______	=	$______
______	______	×	______	=	$______
______	______	×	______	=	$______
			Total monthly savings	=	$______

[a] Either use the values from text table 38 or consult a microwave cookbook to find the times required to cook various foods. Assume that a microwave oven uses 1,500 watts.

TABLE C-78
Expected Savings from Electronic Ignition for Gas Oven or Cook Top

Energy saved = ______ Ccf of gas per month for pilot light(s) (table C-2)

Savings per month = ______ Ccf saved × $______ per Ccf "last" rate (table C-6) = $______

TABLE C-79
Time and Wattage of Dishwasher

Total time = ______ minutes of heater on-time ÷ 60 = ______ hours

Approximation method:

Wattage = ______ watts of heater (from instructions) + ______ watts of motor (from text) = ______ watts

Measurement:

Time for one disc revolution = ______ seconds

Wattage = 3,600 × ______ kh value (table C-1) ______ seconds for one disc revolution = ______ watts

TABLE C-80
Electrical Energy Required to Operate Dishwasher

Energy per use = ______ watts for heater and motor (table C-79) × ______ hours of on-time (table C-79) ÷ 1,000 = ______ kwh

TABLE C-81
Hot Water Used by Dishwasher

Water per cycle = ______ quarts

Total hot water per use	=	______	quarts of water per cycle	× ______	number of wash and rinse cycles used[a]	= ______	quarts
Gallons of hot water per use	=	______	quarts per use	÷ 4 = ______	gallons		

[a] Include the water for any rinse/hold cycles you use.

TABLE C-82
Total Energy Used by Dishwasher

Electricity used per month	=	______	kwh per use (table C-80)	× ______	number of uses per month	= ______	kwh
Hot water used per month	=	______	gallons hot water per use (table C-81)	× ______	number of uses per month	= ______	gallons
Cost of electricity	=	______	kwh used per month	× $______	per kwh average rate (table C-6)	= $______	
Cost of hot water	=	______	gallons used per month	× $______	per gallon average rate (table C-40 or C-41)	= $______	

Total cost per month = $______

TABLE C-83
Hot-Water Fraction of Warm Water Used for Handwashing

Temperature: Hot: ______ °F. Warm: ______ °F. Cold: ______ °F.

Temperature difference between warm and cold water	=	______	°F. warm water	− ______	°F. cold water	= ______	°F.
Temperature difference between hot and cold water	=	______	°F. hot water	− ______	°F. cold water	= ______	°F.
Hot-water fraction	=	______	°F. temperature difference between warm and cold water	÷ ______	°F. temperature difference between hot and cold water	= 0.______	

TABLE C-84
Hot Water Used to Wash Dishes by Hand

Breakfast:

Hot water for wash	=	______	gallons total wash water	×	0.______	hot-water fraction (table C-83)	= ______	gallons hot water
Hot water for rinse	=	______	gallons total rinse water	×	0.______	hot-water fraction (1.0[a] or table C-83)	= ______	gallons hot water

Total hot water used to wash breakfast dishes = ______ gallons

Hot water to wash breakfast dishes per month	=	______	gallons hot water to wash breakfast dishes	×	______	number of breakfast dish washings per month
					= ______	gallons

Lunch:

Hot water for wash	=	______	gallons total wash water	×	0.______	hot-water fraction (table C-83)	= ______	gallons hot water
Hot water for rinse	=	______	gallons total rinse water	×	0.______	hot-water fraction (1.0[a] or table C-83)	= ______	gallons hot water

Total hot water used to wash lunch dishes = ______ gallons

Hot water to wash lunch dishes per month	=	______	gallons hot water to wash lunch dishes	×	______	number of lunch dish washings per month
					= ______	gallons

Dinner:

Hot water for wash	=	______	gallons total wash water	×	0.______	hot-water fraction (table C-83)	= ______	gallons hot water

Hot water for rinse	= ______	gallons total rinse water	× 0.______	hot-water fraction (1.0[a] or table C-83)	= ______	gallons hot water	

Total hot water used to wash dinner dishes = ______ gallons

Hot water to wash dinner dishes per month	= ______	gallons hot water to wash dinner dishes	× ______	number of dinner dish washings per month

= ______ gallons

Total hot water per month for dish washing	= ______	gallons breakfast per month	+ ______	gallons lunch per month	+ ______	gallons dinner per month

= ______ gallons

Cost of hot water per month	= ______	gallons total hot water per month for dish washing	× $______	per gallon average cost (table C-6) = $______

[a] If you use pure hot water for the rinse, use 1.0 instead of the hot-water fraction found for the wash water.

TABLE C-85
Expected Energy Savings from Elimination of One Wash/Rinse Cycle, One Rinse/Hold Cycle, or Both Cycles

Energy saved	= ______	hours on-time of heater during extra wash and rinse	× ______	watts of heater and motor (table C-79)	÷ 1,000

= ______ kwh

Hot water	= ______	quarts per fill (table C-81)	× ______	number of fills for extra wash and rinse	÷ 4 = ______ gallons

Savings for electricity	= ______	kwh saved	× $______	per kwh "last" rate (table C-6) = $______

Savings for hot water	= ______	gallons saved	× $______	per gallon "last" rate (table C-40 or C-41)	= $______	

Total savings per use = $______

Savings per month	= $______	saved per use	× ______	number of uses per month	= $______	

TABLE C-86
Energy Saved by Eliminating Dishwasher Dry Cycle

Energy saved per use	= ______	hours for dry cycle	× ______	watts of heater and motor (table C-79)	÷ 1,000 = ______	kwh
Savings per use	= ______	kwh saved per use	× $______	per kwh "last" rate (table C-6)	= $______	per use
Savings per month	= $______	saved per use	× ______	number of uses per month	= $______	

TABLE C-87
Energy Efficiency Ratio (EER) of Air Conditioner

Wattage:

From your unit: ____________ watts

OR

Rotating disc method:

Time for one disc revolution = ______ seconds

Wattage = 3,600 × ______	Kh value (table C-1)	÷ ______	seconds for one disc revolution	= ______	watts
EER = ______	Btu per hour cooling capacity	÷ ______	watts	= ______	Btu per watt hour

TABLE C-88
Energy Used by Air Conditioning System Last Year

Kwh used during month before
and month after cooling season[a] = ______ kwh + ______ kwh = ______ kwh

Average kwh used per month by other appliances	= ______	kwh used for month before and month after cooling season	÷ 2 = ______	kwh

Month of Cooling Season	*Kwh Used During That Month*		*Average Kwh Used by Other Appliances*		*Kwh for Air Conditioning and Dehumidifier*
______	______	−	______	=	______
______	______	−	______	=	______
______	______	−	______	=	______
______	______	−	______	=	______
______	______	−	______	=	______
______	______	−	______	=	______

Total energy used for air conditioning (and dehumidifier) = ______ kwh

Energy for air conditioning only =	______	kwh for air conditioning (and dehumidifier) −	______	kwh for dehumidifier (table C-95 or C-96) = ______ kwh

Cost per season = ______ total kwh × $______ per kwh average rate (table C-6) = $______

[a] If you have electric heat, be sure to choose months when the heat is not on.

TABLE C-89
Average Electrical Energy Use of Air Conditioner for Past Several Years

Year	*Energy Used for Air Conditioning*[a]
______	______ kwh
______	______ kwh
______	______ kwh
______	______ kwh
______	______ kwh
Total energy =	______ kwh

Average energy for air conditioning per year	= ______	kwh total energy	÷ ______	number of years in above list = ______ kwh per year

[a] Repeat the calculations shown in table C-88 and enter the final value. Include the value for last year in this list.

TABLE C-90
Expected Savings from Increasing Inside Temperature

Change of temperature setting = ______ °F. new setting − ______ °F. old setting = ______ °F.

Fractional savings = ______ °F. change of setting × 0.03 = 0.______

Kwh saved = 0.______ fractional savings × ______ kwh used per year (table C-89) = ______ kwh

Savings = ______ kwh saved × $______ per kwh "last" rate (table C-6) = $______

Expected kwh used after increasing inside temperature = ______ kwh used per year (table C-89) − ______ kwh saved = ______ kwh[a]

---

[a] If you increase the inside temperature and then plan to follow any of the later conservation measures, use this value for your later energy-saving calculations.

TABLE C-91
Expected Savings from Addition of Insulation to Ceiling, Walls, or Both

Ceiling:

Fractional savings = ______ inches of insulation added (9 inches total maximum) × 0.025 = 0.______

Kwh saved = 0.______ fractional savings × ______ kwh used per year[a] (table C-89 or C-90) = ______ kwh

Savings = ______ kwh saved × $______ per kwh "last" rate (table C-6) = $______

Expected kwh used after adding ceiling insulation = ______ kwh used per year[a] (table C-89) or expected kwh used (table C-90) − ______ kwh saved = ______ kwh

Walls:

Kwh saved = ______ kwh used per year[b] (table C-89) or expected kwh used (table C-90 or above) × 0.20 = ______ kwh

Savings = ______ kwh saved × $______ per kwh "last" rate (table C-6) = $______

Expected kwh used after adding wall insulation = ______ kwh used per year[b] (table C-89) or expected kwh used (table C-90 or above) − ______ kwh saved = ______ kwh

[a] If you don't increase the inside temperature, use the value from table C-89; but if you plan to increase the inside temperature, use the value in table C-90.

[b] If you don't plan to increase the inside temperature or add ceiling insulation, use the value from table C-89; but if you increase the inside temperature, add ceiling insulation, or both, use the value from table C-90 (if you only increase the inside temperature) or from the above value (if you add ceiling insulation).

TABLE C-92
Expected Savings from Installation of Storm Windows and Doors

Kwh saved = ______ kwh used per year[a] (table C-89) or expected kwh used (table C-90 or C-91) × 0.08 = ______ kwh

Savings = ______ kwh saved × $______ per kwh "last" rate (table C-6) = $______

[a] Use the value from the last conservation measure that you use.

TABLE C-93
Expected Savings from Purchase of a More Efficient Air Conditioner

Difference between EER values = ______ EER of new unit − ______ EER of old unit (table C-87) = ______

Fractional reduction of energy use = ______ Difference between EER values ÷ ______ EER of new unit = ______

Energy savings per year = ______ kwh used per year[a] (table C-90) or expected kwh used (table C-91, C-92 or C-93) × 0.______ fractional reduction = ______ kwh

Savings per year = ______ kwh saved × $______ per kwh "last" rate (table C-6) = $______

[a] Use the value from the last conservation measure that you use.

TABLE C-94
DEER of a Dehumidifier

Wattage (rotating disc method):

Time for one disc revolution = ______ seconds

Wattage = 3,600 × ______ Kh value (table C-1) ÷ ______ seconds for one disc revolution = ______ watts

Energy used per 24 hours = ______ watts × 24 hours ÷ 1,000 = ______ kwh

DEER = ______ pints per 24-hour capacity[a] ÷ ______ kwh used per 24 hours = ______ pints/kwh

[a] If the capacity is not listed for your dehumidifier, you can find its approximate capacity by measuring the water removed by your dehumidifier during a day when it operates for 24 hours. Temperature and humidity should be close to 80° F. and 60 percent if you want to compare your DEER value to the DEER values of new models.

TABLE C-95
Energy Used by a Dehumidifier: Wattage-Time Method

Month		*Fraction of Time On*		*Wattage (Table C-94)*			*Kwh per Month*
________	:	________	×	________	× 0.732[a]	=	________
________	:	________	×	________	× 0.732	=	________
________	:	________	×	________	× 0.732	=	________
________	:	________	×	________	× 0.732	=	________
________	:	________	×	________	× 0.732	=	________
________	:	________	×	________	× 0.732	=	________
________	:	________	×	________	× 0.732	=	________

Total kwh per year = ________ kwh

Cost per year = ______ kwh used per year × $______ per kwh average rate (table C-6) = $______

[a] This value results from hours per month divided by 1,000 watt hours per kwh.

TABLE C-96
Energy Used by a Dehumidifier: Water Brigade Method

Month		*Pints Removed*		*DEER (table C-94)*		*Kwh Used per Month*
________	:	________	÷	________	=	________
________	:	________	÷	________	=	________
________	:	________	÷	________	=	________
________	:	________	÷	________	=	________
________	:	________	÷	________	=	________
________	:	________	÷	________	=	________
________	:	________	÷	________	=	________

Total energy used per year = ________ kwh

Cost = ______ kwh used per year × $______ per kwh average rate (table C-6) = $______

TABLE C-97
Expected Energy Savings from the Purchase of a New Dehumidifier

New Unit:

Energy used per 24 hours =	______	watts × 24 hours ÷ 1,000 =	______	kwh		
DEER of new unit =	______	pints per 24-hour capacity ÷	______	kwh used per 24 hours =	______	pints/kwh

Difference between DEER values =	______	DEER of new unit −	______	DEER of old unit (Table C-94) =	______	
Fractional reduction of energy use =	______	difference between DEER values ÷	______	DEER of new unit =	0.______	
Energy savings per year =	______	kwh used by old unit per year (table C-95 or C-96) ×	0.______	fractional reduction of energy use =	______	kwh
Savings per year =	______	kwh energy savings per year ×	$______	per kwh "last" rate (table C-6) =	$______	

TABLE C-98
Electrical Energy Used by Clothes Washer Motor

Wattage-time method

Wattage (rotating disc method):

Time for one disc revolution = ______ seconds

Wattage = 3,600 ×	______	Kh value (table C-1) ÷	______	seconds for one disc revolution =	______	watts
Time =	______	minutes total agitation and spin time ÷ 60 =	______	hours		
Energy used =	______	watts ×	______	hours ÷ 1,000 =	______	kwh

Approximation method

Electrical energy = 0.25 kwh

TABLE C-99
Hot Water Used by Clothes Washer

Water to fill tub (from instructions or by measurement):

Regular fill: ______ gallons
Medium fill: ______ gallons
Small fill: ______ gallons

Hot water for one wash and one rinse cycle (no spray rinse):

Hot/warm setting:

Regular wash: ______ gallons for regular fill × 1.5 = ______ gallons
Medium wash: ______ gallons for medium fill × 1.5 = ______ gallons
Small wash: ______ gallons for small fill × 1.5 = ______ gallons

Hot/cold or warm/warm setting:

Regular wash: ______ gallons for regular fill
Medium wash: ______ gallons for medium fill
Small wash: ______ gallons for small fill

Warm/cold setting:

Regular wash: ______ gallons for regular fill × 0.5 = ______ gallons
Medium wash: ______ gallons for medium fill × 0.5 = ______ gallons
Small wash: ______ gallons for small fill × 0.5 = ______ gallons

If your clothes washer has a spray rinse, add 3 gallons to the hot water used for hot/warm and warm/warm settings.[a] The amount of hot water used for a hot/cold or warm/cold setting is the same as that without a spray rinse.

Hot/warm setting (with spray rinse):

Regular wash: ______ gallons for regular wash without spray rinse + 3 gallons[a] = ______ gallons

Medium wash: ______ gallons for medium wash without spray rinse + 3 gallons[a] = ______ gallons

Small wash: ______ gallons for small wash without spray rinse + 3 gallons[a] = ______ gallons

Warm/warm setting (with spray rinse):

Regular wash: ______ gallons for regular wash without spray rinse + 3 gallons[a] = ______ gallons

Medium wash: ______ gallons for medium wash without spray rinse + 3 gallons[a] = ______ gallons

Small wash: ______ gallons for small wash without spray rinse + 3 gallons[a] = ______ gallons

[a] If you measure the amount of water in the spray rinse, use half of that measured amount for the hot-water quantity instead of 3 gallons.

TABLE C-100
Monthly Energy Use of Clothes Washer

Electricity used = ______ kwh per load (table C-98) × ______ number of loads per month = ______ kwh

Gallons of hot water for ______ setting[a] = ______ gallons for ______ setting[a] × ______ number of loads at ______ setting[a] = ______ gallons

Gallons of hot water for ______ setting = ______ gallons for ______ setting × ______ number of loads at ______ setting = ______ gallons

Gallons of hot water for ______ setting = ______ gallons for ______ setting × ______ number of loads at ______ setting = ______ gallons

Gallons of hot water for ______ setting = ______ gallons for ______ setting × ______ number of loads at ______ setting = ______ gallons

Total hot water per month = ______ gallons

Cost of hot water = ______ gallons per month × $______ per gallon average rate (table C-40 or C-41) = $______

Cost of electricity = ______ kwh per month × $______ per kwh average rate (table C-40 or C-41) = $______

Total cost per month = $______

[a] For each different load size and setting, enter the size and temperature settings; for example: H/C Reg., W/C med., and so on.

TABLE C-101
Expected Savings from Changing Washer Rinse from Warm to Cold

Hot water saved per use	= ______	gallons for H/W setting (table C-99)	− ______	gallons for H/C setting (table C-99)	= ______	gallons
Hot water saved per month	= ______	gallons saved per use	× ______	number of warm rinses eliminated per month	= ______	gallons
Savings per month	= ______	gallons saved per month	× $______	per gallon "last" rate (table C-40 or C-41)	= $______	

TABLE C-102
Expected Savings from the Purchase of a New Washer

Hot water saved per load at H/W setting	= ______	gallons presently used for H/W setting	− ______	gallons used by new washer at H/W setting	= ______	gallons
Hot water saved per month	= ______	gallons saved per load at H/W setting	× ______	number of uses per month	= ______	gallons

Hot water saved per month for other settings:[a]

Setting: ____________: ____________ gallons
Setting: ____________: ____________ gallons
Setting: ____________: ____________ gallons

Total saved per month = ______ gallons

Savings per month	= ______	gallons saved per month	× $______	per gallon "last" rate (table C-40 or C-41)	= $______

[a] Repeat the calculation shown for the other settings you use.

TABLE C-103
Energy Required to Dry Laundry: Approximation

Cotton Fabrics:

Weight of laundry = _____ lb. (you plus laundry) − _____ lb. (you only) = _____ lb.

Energy per load = _____ lb. of laundry × {1.7 cf / 0.5 kwh} per lb. = _____ {cf / kwh}

Permanent-press Fabrics:

Weight of laundry = _____ lb. (you plus laundry) − _____ lb. (you only) = _____ lb.

Energy per load = _____ lb. of laundry × {1.1 cf / 0.3 kwh} per lb. = _____ {cf / kwh}

Mixed Loads:[a]

Weight of cotton fabrics = _____ lb.
Weight of permanent-press fabrics = _____ lb.

Energy for cotton = _____ lb. of cotton × {1.7 cf / 0.5 kwh} per lb. = _____ {cf / kwh}

Energy for permanent-press = _____ lb. of permanent-press × {1.1 cf / 0.3 kwh} per lb. = _____ {cf / kwh}

Total energy per load = _____ {cf / kwh}

---

[a] If you mix permanent-press and cotton fabrics in a load (removing the permanent-press when dry ana leaving the cotton fabrics inside to dry longer, measure the weights of the different fabrics. Your mixture may vary from load to load, so try to measure the weights for a "typical" or "average" load.

TABLE C-104
Energy Required to Dry Laundry: Power-Time Method

Wattage or gas flow for different settings:

Setting: ____________________

Electric dryer

Wattage (rotating disc method):

Time for one disc revolution = _____ seconds

Wattage = 3,600 × _____ Kh value (table C-1) ÷ _____ seconds for one disc revolution = _____ watts

Kilowatts = ______ watts ÷ 1,000 = ______ kw

Gas dryer

Cubic feet per hour (plus pilot lights) = 30 ÷ ______ minutes for one revolution of 1/2-cubic-foot dial = ______ cfh

Cfh, dryer only = ______ cfh (dryer plus pilots) − ______ cfh for pilot lights (table C-2) = ______ cfh

Setting:[a] ______________________

Electric dryer: wattage = ______ kw
Gas dryer: gas flow = ______ cfh

Setting:[a] ______________________

Electric dryer: wattage = ______ kw
Gas dryer: gas flow = ______ cfh

Energy per load:

Load: ______________________

Minutes to dry load = ______ minutes total cycle time − ______ minutes to cool-down time = ______ minutes

Hours to dry load = ______ minutes to dry load ÷ 60 = ______ hours

Energy = ______ {cfh / kw} × ______ hours to dry load = ______ {cf / kwh}

Load:[b] ______________________

Energy = ______ {cf / kwh}

Load:[b] ______________________

Energy = ______ {cf / kwh}

---

[a] Repeat the calculations shown for any other setting you use and record the final value here.

[b] Repeat the calculations shown for the other types of loads you dry and record the final value here.

TABLE C-105
Energy Used by Electric Dryer: Meter Reading Method

Type of load: ______________________

Energy per load =	______	meter reading after load is dried[a]	−	______	meter reading before load is dried[a]	= ______ kwh

Type of load: ______________________

Energy per load =	______	meter reading after load is dried	−	______	meter reading before load is dried	= ______ kwh

Type of load: ______________________

Energy per load =	______	meter reading after load is dried	−	______	meter reading before load is dried	= ______ kwh

[a] Try to read the meter to the nearest tenth (0.1) of a kilowatt hour.

TABLE C-106
Energy Used by Gas Dryer: Meter Reading Method

Type of load: ______________________

Cubic feet per load	= ______	number of revolutions of 2-cubic-foot dial	×	2 cf per revolution	= ______ cf

Type of load: ______________________

Cubic feet per load	= ______	number of revolutions of 2-cubic-foot dial	×	2 cf per revolution	= ______ cf

Type of load: ______________________

Cubic feet per load	= ______	number of revolutions of 2-cubic-foot dial	×	2 cf per revolution	= ______ cf

TABLE C-107
Energy and Costs for a Month's Laundry

Type of load: ______________________

Energy per month =	______	{cf / kwh} per load (table C-103, C-104, C-105 or C-106)	× ______	number of loads per month	= ______ {cf / kwh}

Type of load: ____________________

Energy per month = ______ $\left\{\begin{matrix}\text{cf}\\ \text{kwh}\end{matrix}\right\}$ per load × ______ number of loads per month = ______ $\left\{\begin{matrix}\text{cf}\\ \text{kwh}\end{matrix}\right\}$

Type of load: ____________________

Energy per month = ______ $\left\{\begin{matrix}\text{cf}\\ \text{kwh}\end{matrix}\right\}$ per load × ______ number of loads per month = ______ $\left\{\begin{matrix}\text{cf}\\ \text{kwh}\end{matrix}\right\}$

Total energy per month = ______ $\left\{\begin{matrix}\text{cf}\\ \text{kwh}\end{matrix}\right\}$

Ccf of gas per month = ______ cf per month ÷ 100 = ______ Ccf

Cost = ______ $\left\{\begin{matrix}\text{Ccf}\\ \text{kwh}\end{matrix}\right\}$ × $______ per $\left\{\begin{matrix}\text{Ccf}\\ \text{kwh}\end{matrix}\right\}$ average rate = $______
(table C-6)

If you have a gas dryer, add the electrical energy used to tumble the clothes:

Electrical energy to tumble clothes = 0.15 kwh per load × ______ number of loads per month = ______ kwh

Cost = ______ kwh × $______ per kwh average rate = $______
(table C-6)

TABLE C-108
Expected Savings from Purchase of a Dryer Without a Pilot Light

Energy saved = ______ cf of gas used by pilot light per month (table C-2)

Savings per month = ______ cf saved × $______ per Ccf "last" rate ÷ 100 cf per Ccf = $______
(table C-6)

TABLE C-109
Monthly Energy Use of an Iron

Wattage = ______ watts Preheat time[a] = ______ minutes ÷ 60 = ______ hours

Preheat energy = ______ watts × ______ hours preheat ÷ 1,000 = ______ kwh

During ironing:

Time on during one cycle[a] = ______ minutes, ______ seconds
= ______ minutes[b]

Time off during one cycle[a] = ______ minutes, ______ seconds
= ______ minutes[b]

Cycle time	=	______	minutes on during one cycle	+	______	minutes off during one cycle	= ______	minutes

Fraction of time on	=	______	minutes on during one cycle	÷	______	minutes cycle time	= 0.______

Energy for ironing	=	______	watts ×	______	hours ironing time[c] ×	0.______	fraction of time on ÷ 1,000	= ______	kwh

Total energy for one ironing	=	______	kwh for ironing +	______	kwh preheat energy	= ______	kwh

Energy used for ironing per month	=	______	kwh total energy for one ironing ×	______	number of ironings per month	= ______	kwh

Cost per month	=	______	kwh used for ironing per month ×	$______	per kwh average cost (table C-6)	= $______

[a] These times will be different for each different temperature setting you use. Measure these times for the most frequently used setting and assume them to be the times for any setting. The energy differences are not large enough to be worth the effort required to measure the energy for each setting.

[b] Divide the seconds by 60 to convert to fractions of a minute.

[c] Use the average time for one ironing session—don't bother to measure the times for each session.

TABLE C-110
Energy Used by Lights

			Hours Used per Month			*Kwh Used per Month*[a]		
Room	*Fixture*	*Watts*	*Summer*	*Winter*	*Spring/Fall*	*Summer*	*Winter*	*Spring/Fall*
______	______	______	______	______	______	______	______	______
______	______	______	______	______	______	______	______	______
______	______	______	______	______	______	______	______	______
______	______	______	______	______	______	______	______	______
______	______	______	______	______	______	______	______	______
______	______	______	______	______	______	______	______	______
______	______	______	______	______	______	______	______	______
______	______	______	______	______	______	______	______	______
______	______	______	______	______	______	______	______	______
______	______	______	______	______	______	______	______	______
______	______	______	______	______	______	______	______	______
______	______	______	______	______	______	______	______	______
______	______	______	______	______	______	______	______	______
______	______	______	______	______	______	______	______	______
______	______	______	______	______	______	______	______	______
______	______	______	______	______	______	______	______	______
______	______	______	______	______	______	______	______	______
______	______	______	______	______	______	______	______	______
______	______	______	______	______	______	______	______	______
			Total energy used per month =			______	______	______

Summer costs per month = ______ kwh used per month, summer × $______ per kwh average rate (table C-6) = $______

Winter costs per month = ______ kwh used per month, winter × $______ per kwh average rate (table C-6) = $______

Spring/fall costs per month = ______ kwh used per month, spring/fall × $______ per kwh average rate (table C-6) = $______

[a] Multiply the wattage of the fixture by the number of hours used per month during the appropriate season and divide by 1,000. For example, if a 100-watt-bulb is used 30 hours per month during the summer and 40 hours per month during the winter, the summer use is 3 kwh (100 watts × 30 hours ÷ 1,000 = 3 kwh); and winter use is 4 kwh per month.

TABLE C-111
Expected Savings by Reducing On-Times of Lights

Light		*Wattage*		*Reduction of Time per Month*		*Kwh Savings*
______	:	________	×	__________________	÷ 1,000 =	________
______	:	________	×	__________________	÷ 1,000 =	________
______	:	________	×	__________________	÷ 1,000 =	________
______	:	________	×	__________________	÷ 1,000 =	________
______	:	________	×	__________________	÷ 1,000 =	________
______	:	________	×	__________________	÷ 1,000 =	________
______	:	________	×	__________________	÷ 1,000 =	________
______	:	________	×	__________________	÷ 1,000 =	________
______	:	________	×	__________________	÷ 1,000 =	________
______	:	________	×	__________________	÷ 1,000 =	________
					Total saved =	________ kwh

Savings per month = ______ kwh saved × $______ per kwh "last" rate (table C-6) = $______

TABLE C-112
Expected Savings from Decreasing the Wattage of Light Fixtures

Fixture: ________________________

Wattage difference = ______ wattage "before" − ______ wattage "after" = ______ watts

Kwh saved = ______ wattage difference × ______ hours used per month[a] (table C-110) ÷ 1,000 = ______ kwh

Repeat this calculation for all fixtures for which you decrease the wattage, and record the number of kilowatt hours saved in the following spaces.

Fixture	*Kwh Saved*	*Fixture*	*Kwh Saved*
____________	___________	____________	______
____________	___________	____________	______
____________	___________	____________	______
____________	___________	____________	______
____________	___________	____________	______

Total saved per month = ______ kwh

Savings per month = ______ kwh saved per month × $______ per kwh "last" rate (table C-6) = $______

[a] If you plan to reduce the number of on-time hours for lights (table C-111), use the time after reduction.

TABLE C-113
Energy Used by a Television

Wattage of set when operating (rotating disc method):

Time for one disc revolution = ______ seconds

Wattage = 3,600 × ______ Kh value (table C-1) ÷ ______ seconds for one disc revolution = ______ watts

Energy used per month = ______ watts × ______ hours on per month ÷ 1,000 = ______ kwh

If you have an instant-on set, include the following energy:

Wattage of set when off (rotating disc method):

Time for one disc revolution = ______ seconds

Wattage = 3,600 × ______ Kh value (table C-1) ÷ ______ seconds for one disc revolution = ______ watts

Hours off = 732 hours per month − ______ hours on per month = ______ hours

Energy used for off-time = ______ watts when off × ______ hours off ÷ 1,000 = ______ kwh

Total energy per month = ______ kwh used when on per month + ______ kwh per month off = ______ kwh

Repeat the calculations shown for any other television sets you own, and record the kilowatt hours used per month in the following spaces (include the energy for the set already calculated).

Television	*Kwh per Month*
____________	______
____________	______
____________	______

Total used per month = ______ kwh

Cost per month = ______ kwh used per month × $______ per kwh average rate (table C-6) = $______

TABLE C-114
Expected Savings by Turning off Instant-On Television Set

Energy saved
per month = ______ kwh used per month when set is off
(table C-113)

Savings per month = ______ kwh saved per month × $______ per kwh "last" rate = $______
(table C-6)

TABLE C-115
Energy Used by Miscellaneous Appliances

Appliance		*Wattage*		*Hours Used per Month*		*Kwh Used per Month*	*Cost per Month*[a]
______	:	______	×	______	÷ 1,000 =	______	______
______	:	______	×	______	÷ 1,000 =	______	______
______	:	______	×	______	÷ 1,000 =	______	______
______	:	______	×	______	÷ 1,000 =	______	______
______	:	______	×	______	÷ 1,000 =	______	______
______	:	______	×	______	÷ 1,000 =	______	______
______	:	______	×	______	÷ 1,000 =	______	______
______	:	______	×	______	÷ 1,000 =	______	______
______	:	______	×	______	÷ 1,000 =	______	______
______	:	______	×	______	÷ 1,000 =	______	______
					Totals	______ kwh	$______

[a] Cost per month = ______ kwh used per month × $______ per kwh average rate = $______
(table C-6)

TABLE C-116
Expected Savings by Reducing Energy Use of Miscellaneous Appliances

Appliance	*Expected Energy Saved*	*Expected Money Saved*[a]
______	______ kwh	$______
______	______	______
______	______	______
______	______	______
______	______	______
______	______	______
Totals	______ kwh	$______

[a] Savings per month = ______ kwh saved per month × $______ per kwh "last" rate (table C-6) = $______

TABLE C-117
Total Energy Used by All Your Appliances

Appliance	*Oil (gallons)*	*Cost ($)*	*Gas (Ccf)*	*Cost ($)*	*Electricity (kwh)*	*Cost ($)*
Heat (table C-8)	______	______	______	______	______	______
Hot water[a] (table C-45)[b]	______	______	______	______	______	______
Refrigerator, freezer[a] (table C-57, C-58 or C-59)					______	______
Cooking[a] (table C-76)			______	______	______	______
Dishwasher[a,c] (table C-82)					______	______
Air conditioner (table C-89)					______	______
Dehumidifier (table C-95 or C-96)					______	______
Clothes washer[a,c] (table C-100)					______	______
Clothes dryer[a] (table C-107)			______	______	______	______

Iron (table C-109)					______	______
Lights[d] (table C-110)					______	______
Television[a] (table C-113)					______	______
Miscellaneous[a] (table C-115)					______	______
Totals:	______ gallons	$______	______ Ccf	$______	______ kwh	$______

[a] Multiply the monthly use by 12 to determine the annual consumption.

[b] Include the standby heat loss (table C-46, C-47 or C-48).

[c] Do not include hot-water use here; that amount is included in the hot-water total.

[d] Calculate the annual use by adding up the monthly uses, noting the use per month according to the season.

TABLE C-118
Total Expected Savings

Conservation Measure	*Hot Water*		*Oil or Gas*		*Electricity*[a]	
	Gallons	*$*	*Gallons or Ccf*	*$*	*Kwh*	*$*
______	______	______	______	______	______	______
______	______	______	______	______	______	______
______	______	______	______	______	______	______
______	______	______	______	______	______	______
______	______	______	______	______	______	______
______	______	______	______	______	______	______
______	______	______	______	______	______	______
______	______	______	______	______	______	______
______	______	______	______	______	______	______
______	______	______	______	______	______	______
______	______	______	______	______	______	______
______	______	______	______	______	______	______
______	______	______	______	______	______	______
______	______	______	______	______	______	______
______	______	______	______	______	______	______
______	______	______	______	______	______	______
______	______	______	______	______	______	______
______	______	______	______	______	______	______
Totals:	______	______	______	______	______	______

[a] Expected savings are subject to conditions stated in last section of chapter 9.

Appendix D
Degree-days, Heating Days, and Average Outside Temperatures

Table D-1 lists the degree-days, heating days, and average outside temperatures for the cities shown. The first column contains the degree-days, and the next five columns contain the heating days and the average outside temperature (heating days/outside temperature) for the different inside temperatures listed.

The number of heating days for a house with an average inside temperature of 70° F. is the number of days for which the average outside temperature is 65° F. or lower. The other sources of heat in the house (appliances, people, and sunlight) produce enough heat to maintain an inside temperature of 70° when the outside temperature is 65° F. Or, stated another way, these heat sources can produce enough heat to maintain an inside temperature 5° above the outside temperature. If the inside temperature is 68° F., the heating system won't need to supply heat until the average outside temperature falls below 63° F. Therefore the number of heating days is less for maintaining a lower inside temperature, because the heating season starts later in the fall and ends earlier in the spring.

In some regions of the United States, average outside temperature remains in the low 60s during part or all of the summer; so, according to the degree-day concept, the heating system would need to provide heat during part or all of the summer. During the summer months, however, the solar heating will provide most or all of the house heating needs. Thus the maximum heating season is taken to be from September 1 to May 31 (273 days). Even if a house does need some heating during the summer months, the amount is small compared to that used during the rest of the year.

To use table D-1, find the heating days/temperature for the average inside temperature of your house (table C-9 Appendix C). For example, if you live in Phoenix, Arizona, and your house has an average inside temperature of 68° F., the number of heating days is 144 and the average outside temperature during this time is 55° F. For an average inside temperature of 72° F., you would have 168 heating days and an average outside temperature of 57° F.

If you live near none of the cities listed, either use the values for the nearest city or try to interpolate between the values of the two cities nearest to you on either side. Or find the degree-days for your locality (by consulting the nearest weather bureau or referring to the map of degree-days in figure 16), and use the number of heating days and outside temperature of the nearest city that shows about the same number of degree-days in table D-1. For locations in mountainous areas of the country, degree-days, heating days, and average outside temperature can vary greatly over short distances depending on the local geography—so if you live in a mountainous area, be careful how you determine your heating days and average outside temperature.

TABLE D-1
Degree-days, Heating Days, and Average Outside Temperatures for Cities in the U.S.A.

City	*Degree-days*	Heating Days/Average Outside Temperature for an Average Inside Temperature of: *64°F.*	*66°F.*	*68°F.*	*70°F.*	*72°F.*
Alabama:						
Birmingham	2,844	158/48	169/49	180/50	192/51	206/52
Huntsville	3,302	166/47	176/48	188/49	200/50	214/51
Mobile	1,684	153/55	167/56	177/57	190/58	202/59
Montgomery	2,269	143/51	154/52	166/52	179/53	192/54
Alaska:						
Anchorage	10,911	273/28	273/28	273/28	273/28	273/28
Annette	7,053	273/42	273/42	273/42	273/42	273/42
Fairbanks	14,345	273/15	273/15	273/15	273/15	273/15
Juneau	9,007	273/36	273/36	273/36	273/36	273/36
Kodiak	8,860	273/37	273/37	273/37	273/37	273/37
Nome	14,325	273/18	273/18	273/18	273/18	273/18
Arizona:						
Flagstaff	7,322	263/39	273/40	273/40	273/40	273/40
Phoenix	1,552	117/54	131/55	144/55	156/56	168/57
Prescott	4,456	211/45	224/46	236/47	247/47	253/48
Tuscon	1,707	128/54	141/54	153/55	165/56	177/56
Winslow	4,733	203/44	215/45	226/46	238/47	249/48
Yuma	1,005	73/56	95/57	117/58	132/59	145/59
Arkansas:						
Fort Smith	3,336	162/46	172/47	183/48	194/49	209/50
Little Rock	3,354	165/46	175/47	186/48	199/49	213/50
California:						
Bakersfield	2,185	144/52	159/52	173/53	187/54	200/55
Bishop	4,313	203/45	217/46	231/47	247/48	262/49
Blue Canyon	5,704	243/43	249/44	257/44	271/45	273/45
Daggett	1,959	142/51	153/52	165/53	177/53	189/54
Eureka	4,679	273/51	273/51	273/51	273/51	273/51
Fresno	2,650	163/50	176/51	190/52	204/53	219/54
Los Angeles	1,819	144/56	174/57	211/58	224/58	240/59
Mount Shasta	5,890	249/42	257/43	265/44	273/45	273/45
Red Bluff	2,688	162/50	178/51	191/52	203/53	216/54
Sacramento	2,843	163/50	179/51	194/52	212/53	229/54
San Diego	1,507	123/57	156/57	190/58	221/59	233/59
San Francisco	3,042	207/53	227/54	240/54	261/55	273/55
Santa Maria	3,053	216/54	233/54	263/55	273/55	273/55
Stockton	2,806	171/50	186/51	201/52	216/52	231/53
Colorado:						
Alamosa	8,609	273/35	273/35	273/35	273/35	273/35

TABLE D-1
(continued)

City	*Degree-days*	*64°F.*	*66°F.*	*68°F.*	*70°F.*	*72°F.*
Colorado Springs	6,173	249/40	259/41	265/41	273/42	273/42
Denver	6,016	239/41	251/42	259/42	265/42	271/43
Eagle	8,426	273/35	273/35	273/35	273/35	273/35
Grand Junction	5,605	211/40	222/41	233/42	245/42	256/44
Pueblo	5,394	216/41	228/42	240/43	252/44	260/45
Connecticut:						
Bridgeport	5,461	222/41	233/42	246/43	252/44	258/44
Hartford	6,350	233/39	245/40	259/41	265/41	273/42
Delaware:						
Wilmington	4,940	207/42	219/43	230/44	242/45	254/46
District of Columbia:						
National Airport	4,211	187/44	199/45	210/46	222/47	235/48
Dulles Airport	5,010	210/42	222/43	233/44	245/45	257/46
Florida:						
Daytona Beach	897	44/59	85/59	109/59	127/60	144/61
Fort Myers	457	—	—	—	68/64	99/64
Jacksonville	1,237	102/56	120/56	137/57	151/58	166/58
Orlando	704	—	42/61	82/61	105/62	125/63
Pensacola	1,578	118/55	132/55	145/56	157/56	169/57
Tallahassee	1,563	116/55	132/55	145/56	158/56	171/57
Tampa	718	—	41/60	85/61	108/62	129/63
West Palm Beach	299	—	—	—	—	74/66
Georgia:						
Atlanta	3,095	166/48	177/49	189/50	203/51	217/52
Augusta	2,547	153/50	164/51	176/51	188/52	201/53
Columbus	2,378	147/51	157/52	169/52	181/53	193/54
Macon	2,240	142/51	154/51	165/52	177/53	189/53
Rome	3,342	177/46	188/46	199/47	213/48	226/49
Savannah	1,952	132/53	145/53	159/54	172/55	184/55
Idaho:						
Boise	5,833	240/42	253/42	259/43	265/43	270/44
Idaho Falls	8,619	266/34	273/35	273/35	273/35	273/35
Lewiston	5,464	235/43	250/44	257/44	263/45	270/46
Pocatello	7,063	257/38	263/39	268/39	273/40	273/40
Illinois:						
Cairo	3,833	171/44	180/45	191/46	204/47	219/48
Chicago	6,300	225/38	235/39	247/40	258/41	267/42
Moline	6,395	218/37	231/38	243/39	255/41	265/42
Peoria	6,098	214/38	227/39	239/40	251/41	262/42
Rockford	6,845	230/34	242/35	254/36	263/37	269/38
Springfield	5,550	204/39	216/40	229/41	241/43	253/44

TABLE D-1 (continued)

City	*Degree-days*	*64°F.*	*66°F.*	*68°F.*	*70°F.*	*72°F.*
Indiana:						
Evansville	4,624	189/42	201/43	215/44	228/45	241/46
Fort Wayne	6,209	225/39	236/40	248/41	260/42	267/43
Indianapolis	5,577	211/40	223/41	235/42	247/43	259/44
South Bend	6,462	230/38	242/39	253/40	262/41	269/42
Iowa:						
Burlington	6,149	212/37	226/39	239/40	251/41	264/42
Des Moines	6,710	219/36	232/37	246/38	258/40	267/41
Dubuque	7,277	239/36	251/37	263/38	269/38	273/39
Mason City	7,901	245/34	257/35	267/36	273/37	273/37
Sioux City	6,953	222/35	236/36	250/38	261/39	269/40
Waterloo	7,415	236/35	248/36	260/37	269/38	273/38
Kansas:						
Concordia	5,623	204/39	216/40	230/42	243/43	256/44
Dodge City	5,046	200/41	212/42	225/43	239/44	250/45
Goodland	6,119	229/39	241/40	252/41	261/42	266/43
Russell	5,312	199/40	211/41	224/42	236/43	247/44
Topeka	5,243	196/40	209/41	222/42	236/44	248/45
Wichita	4,687	185/41	197/42	211/44	223/45	236/46
Kentucky:						
Lexington	4,729	195/42	207/43	220/44	232/45	245/46
Louisville	4,640	194/42	206/43	219/45	231/46	243/47
Louisiana:						
Alexandria	2,200	143/51	154/52	165/53	178/54	190/54
Baton Rouge	1,670	121/54	135/54	148/55	160/56	171/56
Lake Charles	1,498	112/55	128/55	142/56	154/57	166/57
New Orleans	1,465	109/55	127/56	142/56	155/57	167/58
Shreveport	2,167	138/51	148/52	159/52	170/53	182/54
Maine:						
Caribou	9,632	273/31	273/31	273/31	273/31	273/31
Old Town	8,648	267/34	273/34	273/34	273/34	273/34
Portland	7,498	259/37	266/37	273/38	273/38	273/38
Maryland:						
Baltimore	4,729	202/42	213/43	225/44	236/45	247/46
Massachusetts:						
Boston	6,000	233/40	245/41	256/42	264/43	270/43
Nantucket	5,929	244/42	251/42	259/43	269/44	273/44
Worcester	6,848	246/38	258/39	265/40	273/40	273/40
Michigan:						
Alpena	8,518	268/35	273/35	273/35	273/35	273/35
Detroit	6,325	230/38	240/39	251/40	261/41	269/42
Flint	7,041	245/37	257/38	264/39	273/40	273/40
Grand Rapids	6,801	238/37	250/38	260/39	267/40	273/41
Houghton Lake	8,347	265/35	273/35	273/35	273/35	273/35

TABLE D-1 (continued)

City	*Degree-days*	*64°F.*	*66°F.*	*68°F.*	*70°F.*	*72°F.*
Lansing	6,904	240/37	253/38	263/39	269/40	273/40
Marquette	8,351	265/35	273/36	273/36	273/36	273/36
Muskegon	6,890	245/38	257/39	263/39	270/40	273/40
Sault Ste. Marie	9,193	273/33	273/33	273/33	273/33	273/33
Traverse City	7,698	257/36	264/37	270/37	273/38	273/38
Minnesota:						
Duluth	9,756	273/31	273/31	273/31	273/31	273/31
International Falls	10,547	273/28	273/28	273/28	273/28	273/28
Minneapolis	8,310	251/33	263/34	269/35	273/35	273/35
Rochester	8,227	251/37	262/34	269/35	273/35	273/35
St. Cloud	8,868	259/32	267/32	273/33	273/33	273/33
Mississippi:						
Jackson	2,300	143/51	154/52	165/52	177/53	189/54
Meridian	2,388	146/51	158/51	169/52	180/53	193/54
Missouri:						
Columbia	5,078	194/40	208/42	221/43	234/44	248/45
Kansas City	5,161	194/40	204/41	217/42	230/43	245/45
St. Joseph	5,440	196/39	208/40	221/41	235/43	248/44
St. Louis	4,750	187/41	199/42	212/44	226/45	239/46
Springfield	4,570	187/42	200/43	216/45	229/46	243/47
Montana:						
Billings	7,265	260/38	265/38	269/39	273/39	273/39
Cut Bank	9,033	273/34	273/34	273/34	273/34	273/34
Dillon	8,345	273/36	273/36	273/36	273/36	273/36
Glasgow	8,969	264/32	268/32	273/33	273/33	273/33
Great Falls	7,652	264/37	269/38	273/38	273/38	273/38
Havre	8,853	267/33	273/34	273/34	273/34	273/34
Helena	8,190	268/36	273/36	273/36	273/36	273/36
Kalispell	8,554	273/35	273/35	273/35	273/35	273/35
Lewiston	8,586	273/35	273/35	273/35	273/35	273/35
Miles City	7,889	251/35	262/36	266/36	270/37	273/37
Missoula	7,931	269/37	273/37	273/37	273/37	273/37
Nebraska:						
Grand Island	6,420	221/37	235/39	247/40	259/41	266/42
Lincoln	6,218	213/37	226/39	239/40	251/41	263/42
Norfolk	6,981	227/35	240/37	253/38	263/39	269/40
North Platte	6,743	237/38	249/39	260/40	266/40	270/41
Omaha	6,601	218/36	232/38	246/39	258/40	267/41
Scottsbluff	6,774	246/38	258/39	264/40	269/40	273/41
Valentine	7,300	244/36	255/38	263/40	268/40	273/41
Nevada:						
Elko	7,483	263/38	268/38	273/39	273/39	273/39
Ely	7,814	266/37	273/38	273/38	273/38	273/38
Las Vegas	2,601	161/49	161/50	171/51	181/51	191/52

TABLE D-1 (continued)

	Degree-days	*64°F*	*66°F*	*68°F*	*70°F*	*72°F*
Lovelock	5,990	237/41	251/42	259/43	265/43	272/44
Reno	6,022	254/42	261/43	268/43	273/44	273/44
Tonapah	5,900	237/41	249/42	256/43	263/43	270/44
Winnemucca	6,629	257/40	264/41	270/41	273/42	273/42
New Hampshire:						
Concord	7,360	253/37	263/38	270/38	273/39	273/39
New Jersey:						
Atlantic City	4,946	211/42	223/43	234/44	247/45	257/46
Newark	5,034	208/42	220/43	232/44	243/45	255/46
Trenton	4,947	208/42	221/43	233/44	245/45	257/46
New Mexico:						
Albuquerque	4,292	193/44	205/44	216/45	227/47	239/48
Clayton	5,207	221/42	233/43	247/44	258/45	265/46
Roswell	3,697	178/45	189/46	201/47	214/48	228/49
Truth or Consequences	3,392	174/47	186/48	198/49	211/50	224/51
Tucumcari	4,047	189/45	201/46	214/47	227/48	240/49
Zuni	5,815	237/41	250/42	256/43	264/43	273/44
New York:						
Albany	6,888	238/37	250/38	262/39	268/40	273/41
Binghamton	7,285	250/37	261/38	268/38	273/39	273/39
Buffalo	6,927	245/39	256/40	263/41	270/41	273/42
Massena	8,237	255/34	264/34	273/35	273/35	273/35
New York	5,000	207/42	219/43	231/44	242/45	255/46
Rochester	6,719	239/38	251/39	260/40	268/40	273/41
Syracuse	6,678	238/38	249/39	259/40	266/40	273/41
North Carolina:						
Asheville	4,237	199/45	213/46	228/47	242/48	257/49
Cape Hatteras	2,731	207/51	222/52	235/53	246/53	253/53
Charlotte	3,218	168/47	180/48	192/49	206/50	219/51
Greensboro	3,825	182/45	194/46	207/47	221/48	233/49
Raleigh	3,514	177/46	189/47	202/48	217/49	230/50
Wilmington	2,433	148/50	161/51	174/52	187/53	200/54
North Dakota:						
Bismarck	9,044	263/22	268/32	273/33	273/33	273/33
Fargo	9,271	259/30	267/31	271/31	273/32	273/32
Minot	9,407	266/31	271/31	273/32	273/32	273/32
Williston	9,161	265/31	269/32	273/32	273/32	273/32
Ohio:						
Akron	6,224	229/39	241/40	253/41	262/42	270/42
Cincinnati	4,844	196/42	209/43	221/44	234/45	247/46
Cleveland	6,154	229/39	241/40	255/41	262/42	270/43
Columbus	5,702	218/40	230/41	242/42	254/43	264/44
Dayton	5,641	213/40	226/41	236/43	249/44	260/45
Mansfield	5,818	221/40	233/41	243/42	257/43	263/44

TABLE D-1 (continued)

	Degree-days	*64°F*	*66°F*	*68°F*	*70°F*	*72°F*
Toledo	6,381	229/38	240/39	251/40	262/41	269/42
Youngstown	6,426	236/39	247/40	259/41	266/41	273/42
Oklahoma:						
Oklahoma City	3,695	169/45	179/45	190/46	205/48	221/49
Tulsa	3,680	166/44	177/45	187/46	201/47	217/49
Oregon:						
Astoria	5,295	263/47	273/48	273/48	273/48	273/48
Burns	7,212	261/39	267/39	273/40	273/40	273/40
Eugene	4,739	246/46	253/47	262/47	273/47	273/47
Meacham	7,863	273/38	273/38	273/38	273/38	273/38
Medford	4,930	234/45	247/46	253/46	259/47	268/47
North Bend	4,688	273/50	273/50	273/50	273/50	273/50
Pendleton	5,240	229/43	245/44	255/45	261/45	268/46
Portland	4,792	243/47	253/47	261/48	273/48	273/48
Redmond	6,643	263/41	273/42	273/42	273/42	273/42
Salem	4,852	246/46	253/47	262/47	273/47	273/47
Pennsylvania:						
Allentown	5,827	223/40	235/41	246/42	259/43	266/44
Bradford	7,804	263/37	273/38	273/38	273/38	273/38
Erie	6,851	247/38	256/39	264/40	273/41	273/41
Harrisburg	5,224	208/41	219/42	231/43	243/44	255/45
Philadelphia	4,865	204/42	216/43	227/44	239/45	251/46
Pittsburgh	5,930	226/40	238/41	250/42	262/43	270/44
Wilkes-Barre, Scranton	6,277	231/39	243/40	256/41	265/42	273/43
Williamsport	5,981	224/39	237/40	249/42	261/43	268/43
Rhode Island:						
Block Island	5,771	241/42	247/42	254/43	263/44	273/44
Providence	5,972	236/40	248/41	257/42	263/43	271/43
South Carolina:						
Charleston	2,146	139/52	153/52	166/53	181/54	194/55
Columbia	2,598	151/49	163/50	175/51	187/52	200/53
Greenville-Spartanburg	3,163	168/47	180/48	192/49	205/50	219/51
South Dakota:						
Aberdeen	8,617	254/32	263/33	268/34	273/34	273/34
Huron	8,055	246/33	257/34	265/35	269/35	273/36
Pierre	7,677	241/34	252/35	261/36	266/37	270/37
Rapid City	7,324	253/37	260/38	265/38	270/39	273/39
Sioux Falls	7,838	243/34	254/35	265/36	269/36	273/37
Tennessee:						
Chattanooga	3,505	172/46	183/47	195/48	208/49	222/50
Johnson City	4,306	193/44	206/45	219/46	231/47	244/48
Knoxville	3,478	173/46	184/47	196/48	209/49	222/50
Memphis	3,227	162/47	171/48	183/49	194/50	207/51
Nashville	3,696	173/45	189/46	196/47	209/48	222/49

TABLE D-1 (continued)

	Degree-days	*64°F*	*66°F*	*68°F*	*70°F*	*72°F*
Oak Ridge	3,944	182/45	194/46	208/47	221/48	234/49
Texas:						
Abilene	2,610	147/49	157/50	166/51	177/52	189/52
Amarillo	4,183	187/44	199/45	212/46	226/47	239/48
Austin	1,737	121/53	134/54	145/54	156/55	167/56
Brownsville	650	—	30/60	60/61	90/62	114/63
Corpus Christi	930	56/57	82/58	96/59	113/59	129/60
Dallas	2,290	140/50	150/51	159/51	149/52	160/53
Del Rio	1,523	106/54	121/54	133/55	144/56	154/56
El Paso	2,678	155/49	165/49	176/50	187/51	200/52
Fort Worth	2,380	143/50	152/50	163/51	174/52	186/53
Galveston	1,224	93/56	113/57	129/57	142/58	155/59
Houston	1,434	107/55	123/55	137/56	149/57	161/57
Lubbock	3,545	173/46	184/47	196/48	210/49	225/50
Lufkin	1,940	132/50	144/51	154/51	165/52	176/53
Midland	2,621	149/49	160/50	170/50	180/51	192/52
Port Arthur	1,518	113/55	129/55	143/56	155/57	166/57
San Angelo	2,240	138/51	148/51	158/52	168/53	179/53
San Antonio	1,570	113/54	127/54	140/55	150/56	162/56
Victoria	1,227	91/56	111/56	126/57	140/58	153/58
Waco	2,058	133/51	143/52	153/53	163/53	174/54
Wichita Falls	2,904	152/47	161/48	170/49	180/50	192/51
Utah:						
Bryce Canyon	9,133	273/34	273/34	273/34	273/34	273/34
Cedar City	6,137	241/40	252/41	257/42	264/42	269/43
Milford	6,412	241/39	253/40	258/41	264/41	269/42
Salt Lake City	5,983	230/40	244/41	253/42	259/43	264/43
Wendover	5,760	219/40	230/41	243/42	255/43	260/43
Vermont:						
Burlington	7,876	253/35	264/36	270/36	273/37	273/37
Virginia:						
Lynchburg	4,233	192/44	204/45	218/46	232/47	246/48
Norfolk	3,488	178/47	191/48	204/48	217/49	229/50
Richmond	3,939	184/45	197/46	209/47	222/48	236/49
Roanoke	4,307	195/44	209/45	221/46	235/47	249/48
Washington:						
Olympia	5,530	259/45	273/46	273/46	273/46	273/46
Seattle	5,185	255/46	265/47	273/47	273/47	273/47
Spokane	6,835	256/40	262/40	269/41	273/41	273/41
Walla Walla	4,835	217/44	231/45	248/46	256/47	262/47
Yakima	6,009	240/42	257/44	263/44	271/45	273/45
West Virginia:						
Beckley	5,615	233/42	246/43	258/44	266/44	273/45
Charleston	4,590	197/43	209/44	223/45	237/46	250/47

TABLE D-1 (continued)

	Degree-days	*64°F*	*66°F*	*68°F*	*70°F*	*72°F*
Elkins	5,975	237/41	251/42	263/43	273/44	273/44
Huntington	4,624	197/43	210/44	223/45	236/46	249/47
Parkersburg	4,817	201/41	214/42	226/43	239/44	252/45
Wisconsin:						
Eau Claire	8,388	253/33	264/34	271/35	273/35	273/35
Green Bay	8,098	257/34	266/35	271/36	273/36	273/36
La Crosse	7,417	234/35	248/36	260/37	268/38	273/38
Madison	7,730	251/35	263/36	269/37	273/37	273/37
Milwaukee	7,444	251/36	258/37	265/38	271/38	273/38
Wyoming:						
Casper	7,555	260/37	265/37	269/38	273/38	273/38
Cheyenne	7,255	261/38	267/39	273/39	273/39	273/39
Lander	7,869	260/36	266/36	270/37	273/37	273/37
Rock Springs	8,410	266/35	271/35	273/35	273/35	273/35
Sheridan	7,708	262/37	267/37	271/38	273/38	273/38

Appendix E
Celsius to Fahrenheit Conversions

As this book goes to press, the National Weather Service is in the process of changing temperature scales from Fahrenheit to Celsius degrees. We could have set up all of our calculations to use the Celsius scale, but most home thermostats and household thermometers are the old Fahrenheit models. Since most of the temperatures you measure are in the Fahrenheit scale, we have also retained it for our calculations.

If you use a Celsius thermometer or recent data from the National Weather Service, you must convert Celsius temperatures to Fahrenheit. Table E-1 enables you to convert degrees Celsius to degrees Fahrenheit.

TABLE E-1
Celsius to Fahrenheit Conversion

°C.	°F.	°C.	°F.	°C.	°F.
−10	14	7	45	24	75
− 9	16	8	46	25	77
− 8	18	9	48	26	79
− 7	19	10	50	27	81
− 6	21	11	52	28	82
− 5	23	12	54	29	84
− 4	25	13	55	30	86
− 3	27	14	57	31	88
− 2	28	15	59	32	90
− 1	30	16	61	33	91
0	32	17	63	34	93
1	34	18	64	35	95
2	36	19	66	36	97
3	37	20	68	37	99
4	39	21	70	38	100
5	41	22	72	39	102
6	43	23	73	40	104

Notes

1. Introduction

1. *Chemical and Engineering News,* July 22, 1974, p. 32.

3. Home Heat

1. Btu per hour may be abbreviated as Btu/h.

2. To find the R-value of the total wall area, you must use the conductivity rather than the R-value. Conductivity of the insulated section is 1 ÷ 14.19 = 0.07, and conductivity of the section with studs is 1 ÷ 7.54 = 0.13. Total conductivity of the wall is ⅞ of the conductivity of the insulated section plus ⅛ of the conductivity of the studded section, or 0.062 + 0.016 = .078. The R-value is 1 ÷ conductivity, so the R-value for the wall is 1 ÷ 0.78 = 12.8.

3. Refer to Appendix A for details on how to calculate areas.

4. To convert inches to a fraction of a foot, divide the number of inches by 12: 7 inches is 7/12 = .57 foot; 9 inches = 9/12 = .75 foot, and so on.

5. In this case, round off the R-value of 9.7 to 10 and locate the temperature difference in the 10–12 R-value column.

Useful Literature

Department of Housing and Urban Development. *In the Bank . . . Or Up the Chimney*. Washington D.C.: U.S. Printing Office (20402), 1975 (Stock #023-000-00297-3; cost: $1.70).

Institute on Energy Conservation and the Poor. *Save Energy: Save Money*. Washington D.C.: Office of Economic Opportunity, 1974.

National Mineral Wool Insulation Association, Inc. *How to Save Money by Insulating Your Home*. Summit, N.J.: NMWIA (382 Springfield Ave, 07901; cost: $0.30).

U.S. Department of Commerce. *Making the Most of Your Energy Dollars in Home Heating and Cooling*. Washington D.C.: U.S. Printing Office (20402) (cost: $0.70).

4. Hot-Water Heater

1. Some dishwashers have a preheat cycle during which an electric element heats the water to a prescribed temperature. They save no energy because the electric heating element uses the equivalent amount of energy saved by the hot-water heater. If you have an electric hot-water heater, the hot water for this type of dishwasher will cost no more; but if you have a gas or oil hot-water heater, the electricity used by this dishwasher to heat the water will cost more than the gas or oil that would be used for the same purpose by the hot-water heater. The reduced standby heat loss of your heater, however, will more than make up for this small added cost.

5. Kitchen Appliances

1. The AHAM address is 20 North Wacker Drive, Chicago, Illinois 60606. Price of the *Directory* is 50 cents.

6. Air Conditioners and Dehumidifiers

1. If the unit has no wattage listed but does list the voltage and current, find the wattage by multiplying voltage by current.

2. See Note 1, chapter 5.

Useful Literature

General Services Administration. *Room Air Conditioners*. Washington D.C.: U.S. Printing Office (20402); (cost: $0.45).

Bibliography

3. Home Heat

American Society of Heating, Refrigeration and Air Conditioning Engineers. *ASHRAE 1972 Handbook of Fundamentals*. New York: ASHRAE, 1972.

Anderson, R. W. *Residential Energy Consumption. Single Family Housing*. Columbia, Md: Hittman Associates, National Technical Information Service, 1973.

Eccli, Eugene. *Low-Cost Energy-Efficient Shelter for the Owner and Builder*. Emmaus, Pa.: Rodale Press, 1976.

Federal Energy Administration. *A Pilot Project in Homeowner Energy Conservation*. Washington, D.C.: FEA Office of Communications and Public Affairs, 1974.

"Humidifiers for the Home." *Consumer Reports* 41 (1976): 1017.

Lokmanhekin, M. L., and Anderson, R. W. *Evaluation of Heating Loads in Old Residential Structures*. Columbia, Md.: Hittman Associates, National Technical Information Service, 1974.

NAHB Research Foundation, Inc. *Insulation Manual, Homes, Apartments*. Rockville, Md.: National Association of Home Builders, 1971.

Pennsylvania Power and Light Company. *Insulation Standards and Recommendations*. Allentown, Pa.: PPL, 1977.

Peterson, Stephen R. *Retrofitting Existing Housing for Energy Conservation: An Economic Analysis*. Washington, D.C.: U.S. Government Printing Office, 1974.

Spies, H. R.; Konzo, S.; Calvin, J.; and Thoms, W. *350 Ways to Save Energy (and Money) in Your Home and Car*. New York: Crown Publishers, 1974.

U.S. Weather Bureau. *Climatography of the United States No. 84*. Asheville, N.C.: National Oceanic and Atmospheric Administration.

4. Hot Water Heater

Muller, John George. *The Potential for Energy Savings Through Reductions in Hot Water Consumption*. Washington, D.C.: Federal Energy Administration, 1975.

Pennsylvania Power and Light Company. *A Guide to Reduce Hot Water and Energy Use*. Allentown, Pa.: PPL, 1976.

"Water Heaters." *Consumer Reports* 41 (1976):165.

5. Kitchen Appliances

Association of Home Appliance Manufacturers. *AHAM Report No. HRF-2-ECFT, Standard*. Chicago: AHAM, 1975.

Association of Home Appliance Manufacturers. *1977 Directory of Certified Refrigerators and Freezers*. Chicago: AHAM, 1977.

"Built-in Dishwashers." *Consumer Reports* 41 (1976): 280.

"Gas & Electric Ranges." *Consumer Reports* 39 (1974): 526.

"No-Frost Refrigerators." *Consumer Reports* 40 (1975): 564.

6. Air Conditioners and Dehumidifiers

NAHB Research Foundation, Inc. *Insulation Manual, Homes, Apartments*. Rockville, Md.: National Association of Home Builders, 1971.

Peterson, Stephen R. *Retrofitting Existing Housing for Energy Conservation:An Economic Analysis*. Washington, D.C.: Government Printing Office, 1974.

7. Laundry Appliances

"Clothes Dryers." *Consumer Reports* 39 (1974): 831.

"Washing Machines." *Consumer Reports* (1975): 611.

8. Lights, Television, and All the Other Household Appliances

Weast, Robert C., ed. *Handbook of Chemistry and Physics*, 54th Edition. Cleveland: CRC Press, 1974.

Index